EUROPE IN SPACE

Europe in Space

Guy Collins

Palgrave Macmillan

ISBN 978-1-349-10127-6   ISBN 978-1-349-10125-2 (eBook)
DOI 10.1007/978-1-349-10125-2

First published in the United States of America in 1991

ISBN 978-0-312-05316-1

Library of Congress Cataloging-in-Publication Data
Collins, Guy.
    Europe in space / Guy Collins.
        p.   cm.
    Includes index.
    ISBN 978-0-312-05316-1
    1. Astronautics—Europe—History.   I. Title.
TL789.8.E9C65   1991
338'.0919–dc20   90–44353
                                            CIP

To Anne

Contents

List of Abbreviations   viii

List of Appendices   x

Preface   xi

1   Introduction   1

2   The Pioneering Years   4

3   The Europa Débâcle   15

4   Ariane – From Drawing Board to Launchpad   23

5   Early Scientific Satellites   33

6   Early Communications and Weather Satellites   42

7   Ariane – The Commercial Challenge   49

8   Spacelab   70

9   Remote Sensing, Comets and the Lure of Mars   87

10 The Telecommunication Satellite Revolution   102

11 Satellite Television   110

12 The French Manned Space Programme   123

13 Satellites and Probes in the 1990s   141

14 Shuttles, Space Stations and Politics   153

15 The International Scene and the Way Forward   191

Appendices   201

Notes   216

Index   224

MARECS-A	First of series of maritime communications satellites, based on ECS series and launched in December 1981
METEOSAT-1	First of series of European weather satellites, launched November 1977
MBB	Messerschmitt-Boelkow-Blohm Ag
NASA	National Aeronautics and Space Administration
NASDA	Japanese National Space Development Agency
OTS	Orbital Test Satellite, launched 1978
SAR	Synthetic Aperture Radar
SDI	Strategic Defense Initiative
SEP	Société Européenne de Propulsion
TD-1A	ESRO astronomical satellite launched March 1972
TDF-1	First of pair of French direct broadcast satellites, launched October 1988
TELECOM 1A	First of series of French communications satellites, launched in August 1984
TV-SAT 1	First of pair of West German direct broadcast satellites, launched unsuccessfully November 1987. TV-SAT 2 launched successfully August 1988.

List of Appendices

1. Ariane rocket family   203
2. Ariane 4 rocket series   204
3. Ariane 4 showing industrial contributors   205
4. National contributions to Ariane 3 and 4 development
    programmes   206
5. Ariane industrial programme share-out   207
6. Ariane launch history: the first ten years 1979–89   208–9
7. European Space Agency budget 1988–2000   210
8. Two excerpts from declarations on European space policy
    following meeting of European ministers at The Hague,
    November 1987   211
9. Landmarks in space history   214

Preface

I started writing this book in Paris in the spring of 1987, when the western space industry was passing through a period of trauma and uncertainty. The shuttle had been grounded following the destruction of Challenger in a launch explosion in January 1986, Europe's Ariane rocket had just suffered two failures in only four flights and was also forced to suspend launches, and problems with conventional US rockets meant that by mid-1987 the West had no reliable means of launching payloads of any description into orbit. In contrast the Soviet Union was forging ahead with its long-duration space station flights and launching satellites at the rate of two a week.

Apart from the immediate technical problems there was also a considerable amount of self-doubt clouding western space efforts. In the US NASA was under attack for sloppy management and a lack of credible long-term goals, while in Europe the main industrial nations were divided over whether to move towards a manned space flight programme or whether to concentrate resources on less spectacular unmanned projects. Some research programmes were being delayed by political wrangling, while satellite owners and operators were becoming increasingly frustrated not only at the uncertainty of launch prospects but also the sky-high cost of insurance, which posed a serious threat to the commercial viability of the industry.

By the time I finished the book, in Rome in the summer of 1989, the picture had radically improved. In the United States not only was the shuttle back in action but NASA, inspired by the Ride report, was setting its sights on returning to the moon and perhaps even launching a manned expedition to Mars. The European Space Agency, heartened by the resumption of Ariane launches, had won approval from governments to press ahead with building a mini-space shuttle, Hermes, and collaborating with the US and Japan on constructing an international space station. Despite some unresolved political problems and continuing difficulties over funding, there was a new sense of optimism and urgency in the air.

The book describes the development of the European space programme and analyses the political, financial and technical hurdles which it has faced, as well as assessing future projects and the goals that European nations have set themselves to stay in the space race beyond the turn of the century.

During the course of researching and writing the book I received help from many people associated with the European Space Agency, national space authorities and space industries in various countries. Apart from those mentioned in the text, and many others who helpfully answered questions or pointed me in the right direction, I would particularly like to thank Dr A. Dattner of ESA, who gave assistance in the early stages of research. My thanks go also to Jacqueline Gomérieux of ESA and Claude Sanchez of Arianespace for their help and encouragement during my time in Paris, and to colleagues and friends in the French aerospace journalists' association. Permissions from EAS, HMSO and Arianespace for use of material reproduced in the Appendices is hereby gratefully acknowledged.

I would also like to thank my wife Anne, who displayed enormous patience whenever the writing impinged on family life, as it frequently did. Her good-humoured support made this book possible.

ROME

1 Introduction

When the space shuttle Columbia climbed out over the Atlantic from Cape Canaveral on 28 November 1983 the European space programme, which had been perilously close to collapse a decade earlier, came of age. Riding in the shuttle's cargo bay was Spacelab, an $800 million laboratory designed, built and funded by Western Europe in its first venture into manned research in orbit. Also aboard was West Germany's first astronaut, 42-year-old physicist Ulf Merbold, who, along with the rest of the six-man crew, was to put the laboratory through its paces for the next ten days, delighting mission control by completing an exhausting schedule of experiments with barely a hitch. The flight marked a significant leap forward in Europe's bid to establish itself as a serious space power, confirming Spacelab's versatility and inspiring scientists back on Earth to dream of an eventual permanent European manned presence in space.

The success of the first Spacelab mission was the culmination of more than a quarter of a century of space activity in Europe. Over that period scientists and technicians had built up a wealth of experience in satellite design and construction, and started work on ambitious plans to send probes to the planets and beyond. However, progress in building rockets to launch these vehicles had been slow and arduous, punctuated by a series of set-backs caused by a combination of bad luck, political wrangling, lack of money and poor planning. Perhaps most frustrating of all, even the hard-won successes were always destined to appear modest when set against the spectacular achievements of the US Apollo and Soviet Soyuz manned programmes. Throughout the 1960s and 1970s Western Europe seemed condemned to the status of an also-ran in the headline-grabbing space race between Washington and Moscow.

But while the continent lagged far behind the Americans and Russians in space technology throughout the post-war period, ironically it was the technical genius of German rocket technicians, developed under Hitler during the Second World War, that lay behind the later success of the superpower programmes. The German technicians, who were talent-spotted by the US and Soviet military commands at the end of the war, had started their research during the 1920s and early 1930s and, encouraged by the Nazi war machine, had established undisputed world leadership for German rocketry by 1945. The triumph of the Apollo moon landings of the late 1960s and early 1970s owed much to the European tradition of space science, and the inspiration that Werner von Braun and his team of German rocket scientists brought to the US space effort.

Aside from contributing to the achievements of other nations, Europeans began to realise by the early 1960s that they could also benefit from

1

pursuing an independent space programme of their own. Early efforts were limited mainly to strictly scientific projects, with the scant resources of the time being devoted largely to such activities as sounding rocket programmes and the design of small scientific satellites. As governments and industry grew bolder and more experienced, however, there was a move towards the utilisation of space, first by state organisations supplying public services such as telecommunications or weather forecasting by satellite, and then by private industry interested in the commercialisation of space technology for profit. The change came in a very short time, barely two decades, which meant that people born before the early days of the space era rapidly came to take for granted many aspects of space technology, such as transcontinental satellite telephone links, live television pictures from the other side of the globe and a nightly satellite weather forecast.

The part played by space programmes in providing these services was frequently forgotten. Media coverage of space generally focussed the public's attention on the high-profile manned space flights, pushing the less glamorous space activities into the background. The result was that many people in Europe remained unaware of the scope, and sometimes even the existence, of their country's involvement in space. An exception to the rule was France, where the media since the early 1960s had covered its moves towards establishing an independent space policy with an extraordinary degree of patriotic fervour. But the fact remained that in West Germany, Britain and most of the smaller West European countries, national space policy rarely became a matter of public debate or concern.

While the sums of money involved remained relatively small, and the projects fairly limited, this did not pose a problem. But with the increasing maturity of European space activities, and the growing complexity of the projects, the scale of public debate needed to develop if government and industry were to make informed decisions about the way forward. By the late 1980s, with European governments discussing major spending plans for projects designed to last 20 or 30 years and propel the continent's space programme well into the 21st century, the public ignorance of space programmes was threatening to become a liability. With national governments focussing increasingly on cutting public spending and balancing budgets, pressure was growing on the space industry to justify its projects in strict accounting terms. The notion that space should provide value for money, which was gaining wide acceptance in the 1980s in Britain and the United States, although less so in continental Europe, was a far cry from the brasher 1960s' attitude to space as a last frontier to be conquered at all costs.

The 'value for money' lobby could justly point to the huge demands on public funds from deserving causes such as new hospitals, better educational facilities or even more money for other areas of scientific research.

But those who argued that space should be explored and exploited because it was a challenge that could not be ignored could also point to a history of space projects whose worth had only been demonstrated with hindsight. The revolution in telecommunications brought about by satellites was largely unforeseen in the early days when funds were being committed to experimental communications satellite projects. And even as billions of dollars were being earmarked in the late 1980s for a permanently-manned international space station, scientists were admitting that they had little idea precisely what its scientific value would be. They argued instead that the history of such projects demonstrated that they tended to be ultimately worthwhile, although frequently in unexpected ways.

Space exploration by its nature required both meticulous planning and the capacity to take a leap of faith. Despite the risks and the financial obstacles, European nations by the early 1990s were gradually moving towards building up their own independent space transport capability, establishing a manned presence in space and pushing forward with the exploration, not only of the solar system and deep space but also of the Earth using sophisticated new space-based technology.

Much had been achieved in the first 30 years of the European space programme. But if the self-confidence and maturity that had come to characterise the European space sector by the start of the 1990s justly reflected the scale that its activities had attained, this only served to emphasise how much was owed to the small band of pioneers who had undertaken the original research in the relatively recent past, often on an amateur basis, at a time when space science was much less fashionable and when people who dreamed of building rockets were liable to be dismissed as cranks.

2 The Pioneering Years

GERMAN ROCKETRY UNDER THE NAZIS

One of the leading early rocket pioneers was Hermann Oberth, who had been born in Transylvania, part of the Austro-Hungarian empire, in 1894 but who came to Germany to study physics in the early 1920s. His thesis at the University of Heidelberg, entitled 'Die Rakete zu den Planetenraeumen' (Rockets to interplanetary space), was rejected as being in the realms of fantasy, but he went on to publish a work in 1929 entitled 'Wege zur Raumschiffahrt' (Paths to Space Travel) which became a standard reference text for the early astrophysicists. He became one of the founder members of the German Verein fuer Raumschiffahrt (Society for Space Travel) in 1927, and a year later started working in Berlin as scientific advisor on Fritz Lang's film 'Frau im Mond' (Woman in the Moon).

As part of his contribution to the film he built a small, two-metre high liquid-fuelled rocket. While this rocket never flew, a small rocket motor developed by Oberth dubbed the Kegelduese was ground-tested in July 1930, burning a mixture of liquid oxygen and petrol.[1]

Oberth left Germany in 1930 for his native Transylvania, but other members of the Verein fuer Raumschiffahrt continued his work, notably the young Werner von Braun, a budding rocket scientist who signed a contract of service with the German army in November 1932 to develop military rockets.

Under his army contract he enrolled at the Friedrichs-Wilhelms University and emerged two years later with a doctorate in physics based on the theory and application of liquid-propelled rocket motors. He had already begun to put his theories into practice, building a 150-kilogram rocket called the A1 with a liquid oxygen and ethanol engine more than 40 times as powerful as Oberth's test engine and generating a thrust of close to 300 decanewtons (a newton being the international unit of force required to accelerate a mass of one kilogram by one metre per second per second). In 1934 von Braun's team launched two A2 rockets, a more stable version of the A1, from the island of Borkum in the Baltic Sea, and by the following year they were already testing rockets up to five times as powerful as the A2, working out of an artillery camp at Kummersdorf, near Berlin.[2]

With the rapid growth of the military rocket programme, a more permanent base became necessary, and in April 1936 the Luftwaffe designated the remote area of Peenemuende on the flat North German coast a rocket research centre under von Braun's direction. The move to Peenemuende, and the growing interest of the Luftwaffe in jet engine technology – primarily to power fighter planes rather than rockets – gave von Braun's team of close to a hundred technicians fresh impetus. But the onset of the war in 1939 had mixed implications, making the Peenemuende research more urgent but at the same time giving total priority to the military objectives of rocketry and banishing any immediate dreams of civilian space projects.

The team set itself the task of building a missile which would be able to fire a one-tonne explosive charge over more than 350 kilometres. The missile, at first codenamed the A-4, was to become better known in the latter years of the war in Britain, where the population was on the receiving end, as the V-2. Using revolutionary liquid-fuelled motors which baffled British intelligence and caught the military command across the Channel unawares, the team made rapid progress on the rockets that they hoped would flatten London and swing the war in Germany's favour.

Despite a cut in funding in the early years of the war, and some difficulty in overcoming the German military's scepticism that the rockets would fly in time, the first V-2 was launched in great secrecy from Peenemuende on 3 October 1942 and successfully completed a 118-mile test flight.[3]

In June the following year came the first of the attacks on London by the less-sophisticated V-1 flying bombs, launched from ramps on the French coast, which caused widespread destruction and several thousand deaths in the British capital. But on the night of 17 August the German rocket programme was badly hit when the Peenemuende rocket site came under a night attack from six hundred planes of Britain's bomber command. The raid killed 800 people on the site and forced the research team to split up its operations among several less vulnerable sites further south.

The result was that while the Germans had several thousand V-2s under production at a sprawling subterranean plant at Nordhausen, in the Harz Mountains, within a year of the bombing raid, the missile became operational too late to influence the outcome of the war. While the perfection of the rockets came too late for the Luftwaffe, there was a rapid realisation among the Allied powers of the immense military and scientific advantage that such technology could bring in the peace that followed. This in turn led to a determined effort to track down and recruit the scientists who held the secrets of the German rocket programme.

While many of them, including the two orchestrators of the Peenemuende team, von Braun and General Walter Dornberger, ended up in the United States, others went to the Soviet Union. And a third group were brought to the newly-established French rocket research centre at Vernon, on the Seine northwest of Paris.

Despite the pre-eminence of German rocketry, one of the first engineers in the 20th century to seek to turn the dream of space travel into a reality had been a Frenchman, Robert Esnault-Pelterie. Born in 1881, he had been an aviation pioneer before turning his attentions to space travel. In 1912 he presented a landmark paper to the Société Française de Physique

(French Physics Society) entitled 'Considérations sur les résultats d'un allègement indéfini des moteurs' which studied the theoretical limitations of a rocket whose mass is constantly falling as it burns up fuel. But while he was optimistic that headway could be made in rocketry, he also described the possibility of building a single-stage recoverable space plane using atomic power sometime in the distant future, concluding that such a vehicle would be impossible to build using conventional fuel.

His main legacy to rocket research, apart from his book *L'Astronautique* published in 1930, was his pioneering work on inertial guidance systems, by which rockets could navigate without reference to external data through measuring their own acceleration. While his work in this field was largely theoretical, since the necessary instruments had not been developed, the idea was crucial for the later development of deep-space rockets. Esnault-Pelterie succeeded in building a small rocket motor, with a thrust of some 100 decanewtons, by the late 1930s but lost three fingers in a rocket experiment and had his research career cut short by the outbreak of war.[4]

So it was that the Vernon group of rocket scientists, when they started their work after the war, were working in both a French and German tradition of rocketry that already stretched back more than 30 years. The group, which numbered close to 300 engineers, was recruited under the direction of Rolf Engel, a former SS officer who was co-operating with the French aeronautical research organisation, the Office National d'Etudes et de Recherches Aéronautiques.[5]

ROCKET DEVELOPMENTS IN THE 1950S

Between 1949 and 1957 the team, inspired by Wolfgang Pils, another of the Peenemuende scientists, worked on developing the Véronique rocket probe for the French air force. The first Véronique, a relatively limited rocket designed for high-altitude rather than orbital missions, was launched from the French base at Hammaguir in the Algerian desert in 1952. The rocket, which was produced in several versions, was capable of carrying a 60-kilo package to a height of 210 kilometres (and in its final 1964 version as high as 315 kilometres), and was used for collecting data on the upper atmosphere and near-space.

Launches were conducted through the mid-1950s and in 1957 the French military Comité d'Action Scientifique de la Défense Nationale (CASDN), riding high on the acheivements of the national space effort, decided to mark International Geophysical Year by funding the construction of 15 improved Véronique rockets for enhanced high-altitude research.

Then out of the blue on 5 October 1957 came the announcement from Moscow of the launch the previous day of Sputnik-1, marking the true start of the space age and bringing home to European and American scientists the strength of the competition they faced.

> As a result of intense, large-scale research conducted by Soviet research institutes and design organisations, the world's first artificial satellite ... was successfully launched from the USSR

the statement said.[6]

The event galvanised western space scientists, particularly in the US, and in France focussed political attention on the need to improve co-ordination of the nation's space industry and inject new direction and urgency into the research effort. In the wake of the Sputnik launch scientists in both France and Britain began to appreciate the technological gulf that separated them from their Soviet counterparts.

Research using high-altitude rockets had started slowly in Britain, with the first launches by the Royal Aircraft Establishment of the Skylark rocket probe from the Woomera test range in southern Australia in the mid-1950s. At the same time researchers at De Havilland in Britain had started work in 1955 on the Blue Streak rocket, a single-stage medium-range missile which was intended to form the core of the British nuclear deterrent. But while work was progressing well on the technical level, it became apparent by 1959 that the missile was unlikely to be suitable for military use because it was fuelled with non-storable liquid propellants, notably kerosene and liquid oxygen. This made it difficult to activate quickly, and rendered it cumbersome compared with the solid fuel missiles being developed in the US and France. 'It was a folly, a monument to British self-delusion', wrote Peter Wright, an engineer on the missile's guidance system who subsequently achieved fame as Assistant Director of MI5, in his book *Spycatcher*.[7]

The British military pondered whether to continue with the massive investment entailed by the £1.5 billion project or simply write off the £180 million already spent, and morale among the technicians working on it plummeted.

> We had the capacity ... to put our own satellites into orbit within the following five or six years ... but of course there were those who refused to believe that the projects had any practical or commercial future

Lord Rippon of Hexham, at the time a junior minister in Harold Macmillan's conservative government, recalled many years later.[8]

While Britain was trying to make up its mind what to do, another news item arrived from Moscow to rub salt into the wound. This time Tass announced that a Soviet rocket, Lunik 2, loaded with scientific instruments, had been launched on 12 September 1959 and had successfully travelled the quarter million miles to the moon. It crashed into the lunar surface on 14 September after sending back a wealth of data about the

Earth's magnetic field and meteorite activity en route, and left a small plaque intact on the moon commemorating the flight and proudly affirming its Soviet origins. The development further concentrated minds in Western Europe, and following the cancellation of Blue Streak as a military project in April 1960, was a key factor inspiring the British authorities to propose that the missile should instead form the basis of a jointly-developed civilian European rocket.

FIRST MOVES TOWARDS A EUROPEAN PROGRAMME

The idea that Blue Streak could be more appropriately used in a civilian space programme was first floated by Geoffrey Pardoe, head of the Blue Streak project with De Havilland, during an industrial symposium in late 1959 and was pursued in contacts with French officials during 1960. The early months of the new decade brought few technological developments in Europe but a flurry of organisational and political activity.

The suggestion that Europe needed a science-oriented space organisation had been put forward by Italian physicist Edoardo Amaldi in a letter to a group of other European scientists, including Professor Pierre Auger of France. Auger had been one of the driving forces behind the creation of the European Centre for Nuclear Research (CERN) outside Geneva at the end of the 1950s, which was widely regarded as a model for European scientific co-operation.[9]

And it was hoped that he would impose a similar coherence on the disparate embryonic national space efforts. Amaldi and Auger discussed the idea of a pan-European organisation during a spring walk in the Luxembourg Gardens, Paris, in April 1959, and a more formal discussion of the subject followed in January 1960 at a meeting of the International Council of Scientific Unions' astrophysics committee (COSPAR) in Nice.

> The setting up of a European organisation is an essential and urgent matter if we are not to find ourselves, 20 years hence, confronted by an unbridgeable gulf – scientifically, technically and industrially – separating those countries which can launch spacecraft into interplanetary space from those which cannot

Amaldi wrote in an article in the French-language review 'Expansion Scientifique' published just before the Nice meeting.

A round of meetings in Paris and London during 1960 involving scientists and government officials from various European countries culminated in an 11-nation intergovernmental conference at Meyrin, Switzerland, starting on 28 November that year. The meeting agreed to establish a European Preparatory Conference for Space Research (COPERS) and appointed Auger as its executive secretary.

During 1961 COPERS drew up a relatively ambitious eight-year scientific programme calling for the launch of 440 rocket probes and eight scientific satellites, and laid the groundwork for the creation of two European organisations, one to handle rocket developments and the other the scientific programmes. A major step was taken in January of that year when the French government agreed to pool its research effort with Britain in the interests of building a European satellite launch vehicle using Blue Streak as the first stage, despite President Charles de Gaulle's legendary disdain for co-operation with the British.[10]

Later the same month Britain and France issued an invitation to other European countries to join the project. The result was a blueprint for a 31.7 metre rocket with a launch weight of 104 tonnes using Blue Streak as the first stage, a French second stage called Coralie and a small third stage to be built by West Germany. The rocket, which was the first example of European co-operation on a space project, was given the somewhat grandiose name Europa 1.

Following the decision to press ahead with an experimental heavy-lift launcher, preparations were completed to set up an organisation to supervise the programme. And on 29 March 1962 a convention setting up the European Launcher Development Organisation (ELDO) was formally signed in London by its seven member states – France, Britain, West Germany, Italy, Belgium, the Netherlands and Australia.

Australia owed its membership to the rocket launch site at Woomera, which the British had used since the mid-1950s, while Denmark and Switzerland took observer status. ELDO's charter defined the aims of the organisation as peaceful, and independent of any European military programmes, despite its reliance on years of military-funded research which had contributed to the production of the Blue Streak prototype. Its headquarters were set up in Neuilly, in western Paris, with a very small initial staff of just 53 in 1962, which grew to 130 by the time the organisation formally started operating in February 1964.

Meanwhile parallel moves continued towards setting up an organisation devoted to scientific space programmes and the construction of satellites. The convention establishing the European Space Research Organisation (ESRO) was signed on 14 June 1962 and nine countries became full members – the six European members of ELDO plus Spain, Sweden and Switzerland, with Denmark becoming the tenth member in December 1962. As with ELDO, ratification dragged on, and the organisation did not formally start operating until 20 March 1964.

EUROPEAN SPACE RESEARCH ORGANISATION (ESRO)

ESRO was intended from the start to function independently of ELDO, designing sounding rocket experiments and satellite projects but taking no part in the development of launch vehicles. As a result, when the Europa rocket started to run into serious problems in the mid-1960s, ESRO was able to continue its programmes through launches offered by the US National Aeronautics and Space Administration (NASA). ESRO was given a $306 million budget to cover its first eight-year operational period 1964–72, and set to work to try to recoup some of the ground lost by the lack of progress in the years since the Sputnik launch.

One of its first actions was to set up the European Space Research and Technology Centre (ESTEC) to co-ordinate research. For political reasons it was agreed early on that the various European space installations should be spread out among several countries, and ESTEC was installed provisionally in an austere building provided by the Technical University in the Dutch town of Delft, moving subsequently to a site on the sand dunes at Noordwijk, on the wind-swept North Sea coast a few kilometres north of The Hague. Scientists started work in a group of prefabricated huts close to the beach while permanent buildings were being constructed.

Apart from ESTEC, several other bodies were set up in the mid-1960s under the ESRO umbrella including a satellite tracking and data processing centre (ESDAC) at Darmstadt, West Germany, a space physics research laboratory (ESRIN) at Frascati near Rome and a tracking and telemetry network (ESTRACK) comprising stations in Alaska, the Falkland Islands, Spitzbergen and Belgium.

EUROPEAN LAUNCHER DEVELOPMENT ORGANISATION (ELDO)

In parallel with the establishment of the ESRO structure and facilities, work was progressing on setting up ELDO. As a rocket organisation ELDO had a far higher political profile than ESRO, since it was more accessible to the public and invited comparison with the exploits of the US and Soviet Union.

The Soviet success in launching Yuri Gagarin as the first man in space aboard Vostok 1 on 12 April 1961, followed by John Glenn's feat in becoming the first American astronaut in orbit in February 1962, both predated the signing of the ELDO convention by European states. This meant that at the time when Europe was setting itself the goal of building a rocket capable of launching satellites into orbit by 1967, the European public was already becoming accustomed to newspaper reports and pictures of manned space flight.

In contrast, the original goal of ELDO appeared modest – to build the Europa 1 rocket which would be able to launch a one-tonne satellite into a circular orbit at an altitude of about 300 kilometres. The initial cost of the programme, spread over five years, was estimated at £70 million, or some US $210 million at 1962 exchange rates.

An immediate question which faced delegates at the Lancaster House conference in October 1961 was whether ELDO should devote all its resources to building a rocket based on a 1950s British design – albeit the most powerful in Europe at the time – or whether it should spend more on research into more up-to-date technology. According to an ELDO report to the Council of Europe, a committee was formed to study the problem.

This body arrived at a synthesis between the views of those delegations which advocated the development of a rocket having Blue Streak as a first stage, using the techniques already mastered . . . and the concepts, mainly canvassed by the Italian Delegation, which argued that Europe should as early as possible tackle more advanced techniques.[11]

It was agreed to concentrate on the development of Europa 1, based on the Blue Streak first stage, as planned but to allocate £2 million to research on more advanced technology at the same time. It had been agreed in January 1961 that the Blue Streak first stage would be powered by British Rolls-Royce engines while France and West Germany would build the smaller second and third stages. ELDO, aware that it would have to fight hard to fund the programme, stressed the industrial spin-offs to be expected.

The development and construction of launcher systems . . . together with the construction of satellites themselves require new capabilities in technology

ELDO president Professor G. Bock wrote.

These advances provide new stimuli and incentives in many other fields of activity among which are electronics, materials production and processing, industrial chemistry and precision engineering . . . The participation of the countries of Europe in space activities will therefore promote not only their scientific knowledge but also the progress of European industry as a whole.[12]

But European governments were well aware that while they might hope for technological spin-offs fairly rapidly for their industries, it would be many years before they could hope to start recovering any of the investment directly from space activities.

Discussions on funding Europa 1 development quickly ran into problems since the original proposal that each country should contribute in proportion to its national income proved unworkable. This was because British firms involved with the Blue Streak first stage would inevitably win a disproportionate share of the work. As a result it was agreed that Britain, which would have funded around one quarter of ELDO's budget on a national income basis, would in fact pay a third, with France and West Germany paying around one-fifth each and the smaller countries supplying the rest.

The original timetable called for a ten-launch development programme, with the complete rocket flying by 1965. But the delays in starting the programme, and technical improvements to the third stage, led to a radical review of the costs and timetable in 1964 and early 1965. At the same time France was becoming concerned that Europa 1 was already techically outdated before it had even flown, and demanded a drastic rethink of its design to enable it to launch telecommunications satellites.

THE REVISED EUROPA PROGRAMME

An intergovernmental meeting in Paris in January 1965 concluded that a more realistic estimate of the cost of completing the Europa 1 five-year development programme would be at least $300 million, 50 per cent more than the initial estimate. The meeting also set up a working group which fixed as a new objective the construction of a rocket called ELDO-B, maintaining Blue Streak as the first stage with one or two additional stages powered by liquid hydrogen. Three months later the ELDO conference adopted a two-pronged strategy. Work would start on ELDO-B, transforming the Europa 1 into a rocket to launch telecommunications satellites, but would also continue on the original research programme, for fear that suspending it with half the work done could prove costly later if teams of scientists and researchers needed to be reassembled.

As a result a new launch timetable was drawn up. Three flights of the Blue Streak by itself and with an inactive upper stage – the least ambitious part of the programme – had already been successfully conducted between June 1964 and March 1965, but engineers knew that the hardest part still lay ahead. And they could only begin to guess at the political minefield which the project would face in the years to come.

The second phase of the development programme was to test the flight characteristics of the complete Europa rocket, initially using dummy upper stages but with live separation systems. Apart from testing the rocket design, this would also put the launch facilities at Woomera through their paces. However, the delays in setting up ELDO and arising from the redesign of the rocket meant that the target date for completing test flights was running some four years behind schedule, with the final test launch, originally scheduled for 1965, now targetted for 1969.

This serious slippage in European ambitions occurred against the background of continuing well-publicised feats of manned space flight by both the Soviet Union and the United States. President John F. Kennedy had already set the ambitious target for NASA of putting a man on the moon by the end of the decade. On 18 March 1965 the Soviet Union passed another milestone when Alexei Leonov squeezed through the airlock of his Voskhod 2 spacecraft to become the first man to walk in space. Although he spent barely 20 minutes outside the capsule, he demonstrated that it was possible to survive and work in that extremely hostile environment, characterised by total vacuum, violent temperature swings and radiation. Less than three months later Ed White performed the same feat aboard the US spaceship Gemini 4, while in December that year the US clocked up another space first when Gemini 6 performed a rendezvous with the orbiting Gemini 7, a manoeuvre which was a stepping stone towards the later moon missions.

The delays and wrangling that plagued the European rocket programme made a depressing contrast but the mood was not all gloom, for despite the problems with Europa, France had been working independently on a national rocket programme of its own.

THE DIAMANT PROGRAMME

As early as 1957, before the Sputnik launch, a team of French scientists went to the United States as part of a study to see if it was feasible to convert ballistic missiles that France was developing for its strategic deterrent into satellite launchers. The proposal was to convert one of the missiles, codenamed Saphir, Topaze and Emeraude, into a rocket capable of launching a 50- to 100-kilo satellite into a 300-kilometre high orbit. The study was encouraging, and in December 1960 the state rocket company Société pour l'Etude et la Réalisation d'Engins Balistiques (SEREB) submitted its report to the French defence ministry.

> It is possible to build, at modest extra cost to that of the work already started on experimental surface-to-surface ballistic systems, a satellite launcher capable of placing a mass of 50 kilograms with acceptable precision in an orbit around the Earth with a perigee of 300 kilometres (or 60 kg with a 200 kilometre perigee). It would additionally be possible to build two more developed versions, the first capable of launching an 80 kg satellite and the second a 100 kg satellite into a 300 kilometre perigee orbit. The development programme currently envisaged would permit the completion of the first version in mid-1964, the second version in mid-1965 and the third at the start of 1966. [13]

In December 1961 the government decided to build the second of the three proposed versions, designated Diamant (diamond), and on 9 May 1962 the state arms procurement agency DMA signed an accord with the French space agency CNES making SEREB responsible for building the rocket, DMA for ensuring that it met technical specifications and CNES for financing the project.

Over the following three and a half years France launched 38 rockets in the 'precious stone' series from its base at Hammaguir in the Algerian desert. The first Emeraude was launched on 17 June 1964, a slim, 18-metre rocket designed as the first stage of the Diamant-A, while in July 1965 the first Saphir was launched, comprising an Emeraude with a Topaze second stage.

At this point the civilian and military programmes were still running in tandem, for Saphir had two objectives – to test the first and second stage rocket and guidance systems of the Diamant, and to provide information on the re-entry characteristics of nuclear warheads into the atmosphere. Finally technicians were satisfied that the Diamant-A was ready to go. And on the afternoon of 26 November 1965 the rocket rose from the Hammaguir launchpad into the clear blue Saharan sky, carrying in its nose-cone the first French test satellite Astérix, a tiny 42-kilo package whose role was to confirm the rocket's ability to place it in orbit. The event marked a landmark for France, making it only the third country to launch a satellite and signalling further progress in its quest for an independent role on the global stage. It also drove home to other European countries the leading role that France would have to play if the continent was to develop its own supranational launch capability.

3 The Europa Débâcle

ELDO's attempts to give Europe an independent satellite launch capability in the 1960s and early 1970s were severely hampered by a series of technical and political setbacks. But in addition the organisation suffered from the start from a divergence of motives among member countries, and an unclear perception of the way satellite technology would develop. This lack of direction was responsible for the original design of Europa I as a rocket capable only of launching payloads into low Earth orbit – useless for telecommunications satellites which were to prove the fastest-growing sector of the space market.

BRITISH DISILLUSION – ELDO'S POLITICAL CRISIS

The switch to the Europa II design, at the instigation of France, at least gave the proposed rocket a geostationary capability which could assure it a more useful future. However, it did not resolve some fundamental worries over European space goals, and these resurfaced dramatically in February 1966 when the British Aviation Minister, Fred Mulley, circulated a memorandum to the other ELDO members casting doubt on the entire value of the organisation. A flurry of top-level meetings ensued in April, June and July of that year, as governments sought to re-establish confidence in Europe's embryonic space venture.

The inherent weakness of the ELDO structure was identified by General R. Aubinière, its council chairman from 1968 to 1970 and secretary-general from 1972 to 1973, in an article he wrote marking the tenth anniversary of ELDO in March 1974:

> The idea of creating a group of European countries to build a rocket capable of placing heavy satellites in low orbit ... was inspired more by the need to find some use for a rocket already wholly developed by one country – namely the United Kingdom's Blue Streak, which had just lost its military raison d'être – than by any real resolve to make Europe self-sufficient in the satellite launcher area. What is more, the European countries that finally became partners in the venture agreed to share the tasks under the responsibility of their respective Governments, leaving to the international organisation they were setting up only very limited powers in respect of technical and financial management of the project.[1]

The review of ELDO's structure led to a series of formal decisions in July 1966, notably turning the ad hoc ELDO ministerial conference into a full-time body, the European Space Conference, meeting annually. A smaller group of ministers was set up to improve co-ordination of European space activity, while ELDO itself was given more power to hand out contracts autonomously.

ELDO members agreed to press on with the Europa I rocket programme, despite the technical setbacks, with continental European countries digging deeper into their pockets to finance a cut in the British contribution to 27 per cent from 38 per cent. And the go-ahead for the Europa II project was confirmed, which space officials hoped would give Europe the relatively modest capability of placing 200 kilogram satellites into geostationary orbit through the simple expedient of adding two upper stages, built by France and Italy, to the Europa I.

European research ministers met in Paris in December 1966 to confirm the decisions of one of the most turbulent years to date in the European rocket programme. Yet within months the Europa programme appeared to be veering seriously off course again.

Two launch failures in August and December 1967, due to problems with the French second stage Coralie, called the design of the whole project into question. Then in April 1968 a political bombshell struck when the British government decided the ELDO rocket programmes had no convincing economic future, and announced that it would stop funding the venture after 1971, when its existing commitments came to an end. Lord Rippon of Hexham recalled later:

> We remained in the forefront of development until the Labour Government ... abandoned Blue Streak, shut down the facility at Spadeadam, which was one of the most advanced installations in the world, and turned away 900 skilled people working there. Far from keeping the pledge to forge the new Britain in the white heat of the technological revolution, all they left was a flickering ember.[2]

The response from other European countries was to attempt to salvage what they could from a programme which was clearly heading for the rocks, both politically and technically. After a series of ministerial-level meetings during the year, European governments agreed to scale back the Europa I and II programmes by cancelling the final test launch and Italian booster stage, in a bid to keep the project within the 1966 budget. The agreement coincided with the failure of the F7 launch in November 1968, due to a malfunction in the German third stage Astris.

By the end of the year a split had developed between the countries still enthusiastically backing a European rocket programme, led by France and including West Germany, the Netherlands and Belgium, and those which had become disillusioned with the whole project, namely Britain and Italy.

In December of that year the British and Italian governments formally served notice that they were not interested in participating in the future of the programme, a dramatic reversal of Britain's initial role at the start of the decade as chief inspiration and funder of the project. As a result the British said that they were prepared to fund only half of the spending up to 1971 to which they were committed under the 1966 agreement. This created a further budgetary crisis which was not resolved until April 1969 when France, West Germany, Belgium and the Netherlands agreed to make up the difference themselves to keep the Europa programme alive.

Just as the political crisis appeared to have been patched up there was more bad news on the technical front. On 3 July 1969 another attempt to launch the Europa I failed, again because of the non-ignition of the German-built third stage. While reinforcing the gloom among those connected with the project, the event attracted virtually no attention in Europe outside the space community.

Instead public interest was focussed on the Apollo 11 moon landing that month, dramatic evidence of the huge technological lead which the US space programme had established not only over the rival Soviet programme, but over the still embryonic, and faltering, West European efforts. The giant 110-metre 2700 tonne Saturn 5 rocket which took Neil Armstrong, Buzz Aldrin and Michael Collins to the moon dwarfed all European efforts to date.

And when, on the evening of 20 July, NASA's lunar module settled into the powdery dust of the moon's surface and Armstrong's voice crackled out over a quarter of a million miles of space with the message 'Tranquillity Base here, the Eagle has landed', the event served both as an inspiration for what could be achieved in space, and a reminder of how little progress Europe had made so far in the field.

EUROPA III

Foundation stones were being laid in Europe, however. At the crisis meeting of ELDO ministers in April 1969 the go-ahead had been given for studies on yet a third version of the Europa rocket, the Europa III B. With the development of satellite technology, the inadequacies of the Europa II launcher, which could theoretically place a maximum of only 200 kilograms into geostationary orbit, were becoming more and more apparent, even ignoring the string of launch failures involving its precursor, Europa I.

Ministers agreed that the new rocket should be capable of launching 750 kilogram satellites into geostationary orbit, more in line with the size of telecommunications satellites already on the drawing board for the 1970s.

Design work started in the following months, and in May 1970 ministers approved a plan for a two-stage rocket. The first stage, powered by four French Viking motors burning 150 tonnes of fuel, was designed to take it up through the Earth's atmosphere, and the second cryogenic stage, fuelled by 20 tonnes of liquid oxygen and hydrogen, would ignite in space. The motor design selected to power the second stage was the H20 proposed by France's Société Européenne de Propulsion (SEP) and developed by a group called Cryorocket, formed jointly by SEP and Messerschmitt-Boelkow-Blohm of West Germany.

Once again, however, the momentum generated by progress on the Europa III studies and the political support for the rocket suffered a blow with the failure of the tenth Europa test on 12 June 1970. As it was the final Europa I test-launch from the Woomera site, hopes had been riding high that it would mark a change in fortunes from the series of failures which had characterised the previous three years. This time, however, while the first and second stages worked well and the German third stage ignited, it did not perform to full strength and the satellite protection cone failed to separate. As a result the ten launches of Blue Streak and Europa I from Woomera, spanning the six years June 1964 to June 1970, came to an end without a single satellite having been placed in orbit.

ESRO/NASA TALKS

Against the background of the Europa programme's tortuous development, talks had started between European space scientists and the US National Aeronautics and Space Administration (NASA) to explore ways of co-operating on joint projects. These talks, led by Sir Hermann Bondi, the director-general of the European Space Research Organisation (ESRO), focussed on several issues, including the possibility of Europe renouncing its rocket programme and turning instead to a reliance on US launchers.

NASA came up with a proposal that Europe should build a space tug, a vehicle designed to take satellites from the low Earth orbit into which the planned US space shuttle would place them to the higher orbit they needed to reach in order to function. This proposal sounded attractive to some countries, notably West Germany, since it ensured a degree of mutual dependency – Europe relying on NASA to provide the transport to low-Earth orbit, but NASA relying on Europe to place the satellites in their final positions.

> There is no doubt that full-scale cooperation with NASA on these great projects could give a tremendous impetus to the European space effort. European industry would be working on a large scale in the very forefront of space technology

the 1969 ESRO report commented enthusiastically.[3]

The May 1970 meeting of ELDO ministers gave the green light for studies on the space tug concept, and negotiations with both NASA and the US State Department continued over the following two years to define the conditions of European access to American rockets, and the question of financing.

However, disagreement surfaced over a US declaration of its right to place conditions on the launch of European satellites, calling into question the whole issue of European access to space. The dispute foreshadowed a similar difference of outlook which would lead to tough negotiations 15 years later over management of a proposed international space station, and the discussions were finally called off when NASA revised its original offer in June 1972. Instead of Europe building the tug, an integral part of the satellite launch system, NASA suggested that it should build a space laboratory which would ride in the shuttle's cargo bay.

> When it became clear that the only item remaining for possible European participation in the USA's post-Apollo programme was the space laboratory, all work sponsored by ELDO on studies on the space tug and on Shuttle technology was brought to an end ... the last meeting of the ELDO/NASA tug steering committee took place in October 1972

a European Space Agency report reviewing the period 1964/84, said.[4]

While the proposal to build a laboratory was subsequently taken up, resulting in the Spacelab programme of the early 1980s, the abortive outcome of the talks with NASA on launch co-operation undoubtedly had a dispiriting effect on the ELDO administration.

THE EUROPA II LAUNCH

In the intervening period ELDO had suffered another blow, this time one that was to prove crippling. On 5 November 1971 the first – and as it was to prove, the last – Europea II rocket was launched from the new French space centre at Kourou, French Guiana.

After a fanfare of publicity, and in the presence of a large number of VIPs and journalists, the rocket lifted off from the pad, only to explode two and a half minutes into its flight after an electrical discharge had caused a breakdown in the inertial guidance system, causing the rocket to veer off course.

> The Europa II rocket plunged calamitously into the Atlantic today, bringing dismay to the four European countries which had built it.

Times correspondent Pearce Wright wrote in a report from Kourou.

> Failure of a rocket costing more than £4 million is bad enough, but a great deal more hung on the outcome of this particular launching. There has been growing disillusion about the continuation of an organisation which has spent over £250 million – particularly when launch vehicles

can be bought from the Americans at £2 million to £3 million each ...
The French and Germans are committed to an agreement for launching
their own European communications satellite system ... but other
members of ELDO are considering leaving unless there are more
obvious signs of progress. Today's launch was intended to provide this
boost to morale. In practice it has probably sounded the death knell for
the organisation as it is constituted.[5]

The message was not lost on those most closely involved in the
programme. Heinz Kaminski, director of the West German space research
institute in Bochum, reacted sharply to the launch failure, pinning the
blame for technical problems on poor organisation and over-politicisation
of the programme. He urged the formation of a European space authority
along supranational and purely scientific lines to overcome these problems.
    And General Aubinière, who was to take over as secretary-general of
ELDO at the start of 1972, wrote two years later that

the failure of F11 in November 1971 brought home to the member states
– and this was indeed the only positive point it achieved – the necessity
for a complete overhaul of programme management methods.[6]

Aubinière chaired a commission of enquiry into the F11 failure and the
wider problems of the programme, which reported in May 1972. The
enquiry concluded that there were a number of interlinked problems,
including a lack of homogeneity between the stages of the rocket, each
designed by a different country. The report, among other points, was
highly critical of the German third stage, which had caused the failure of
the F8 and F9 launches, citing serious design faults in its guidance and
telemetry systems.

FRENCH DIPLOMACY AND FIRST MOVES TOWARDS ARIANE

The French government, which was becoming increasingly impatient with
the lack of progress in building a rocket that would work, forced the pace
through the summer of 1972. At a Franco-German summit meeting in
Bonn in July 1972, French President Georges Pompidou urged West
German Chancellor Willy Brandt to throw his country's weight behind the
development of the Europa III programme, and to be wary of US
overtures on space co-operation. West Germany was interested in the
Spacelab project, however, and reluctant to scale down links with NASA
under political pressure from across the Rhine.
    While consultations continued between France, Britain and West Ger-
many over the way forward, the French national space agency (CNES) was
starting work on a project to replace the whole Europa rocket concept with

a new, technologically simpler launcher. CNES finalised a report on the rocket, dubbed the L-3S (launcher with three substitute stages), on 19 September 1972 and on 20 December submitted the proposal to ministers at the fifth European Space Conference in Brussels.

The conference proved to be a landmark in the development of the European space programme. Not only did it approve the L-3S launcher, designating France project leader, but it also gave the green light for the development of Spacelab under West German leadership. And on the organisational level it finally decided to merge the parallel rocket activities of ELDO and satellite development and scientific research being carried out by ESRO under the umbrella of a single organisation, the new European Space Agency.

As a result of the adoption of the new, simplified L-3S rocket programme, the Europa III project was definitively dropped. However, the future of the Europa II, which had only had one test launch – the November 1972 failure – remained unclear for a few months, until West Germany made clear at an ELDO council meeting in April 1973 that it saw no point in continuing. The second test Europa II was actually en route to French Guiana for launch when the decision to scrap the programme was made.

With the setting up of the European Space Agency, most of ELDO's 530 staff were made redundant in 1973, with the exception of around 60 who were given temporary contracts with ESRO, which had taken over responsibility for the L-3S project until the new ESA could be set up. ELDO's industrial contracts were wound up during 1973, and most of its hardware was also handed over temporarily to ESRO.

ELDO secretary-general Aubinière, reviewing his organisation's turbulent history, identified two main reasons for its failure to achieve its relatively limited objective – to build a rocket that could place a satellite in orbit.

> First and foremost, projects of the magnitude and complexity of a space launcher are not to be improvised, and no initial mistake can ever be put right later

he wrote of the technical problems plaguing the successive launch attempts. And turning to the chequered political support for the organisation through the years, he commented that

> the essential is that this kind of project should answer a need acknowledged by all the member states, who will therefore be prepared with a common political determination to see it through to completion. Such was unfortunately not the case for the Europa programmes, and there we have the underlying cause of ELDO's failure.[7]

If 1972 was a year for dwelling on the past, when the full extent of the Europa mistakes was finally brought home to the ELDO governments,

1973 was the year when Europe turned the page and began to look forward
to the new ventures – L-3S, Spacelab and the creation of the European
Space Agency. One symbolic, if light-hearted, moment came on 1 August
1973, during a meeting of the European space conference in Brussels,
when it was suggested that a more inspiring name should be found for the
L-3S project.

The round-table discussion produced a host of suggestions, ranging from
classical names such as Prometheus and Argo, through Phoenix, Egmont
and Venus to stranger proposals such as Crab and even Edelweiss, to mark
the fact that 1 August was the Swiss National Holiday. In the end it was not
until later in the year that a decision was made, settling for the French-
inspired choice Ariane. In classical mythology it was Ariane's thread which
helped Theseus out of a maze, and the European space community was
hoping that the new rocket would mark a turning point which would help it
out of the maze into which it too had strayed.

In September 1973 the Spacelab Agreement between the US and
Europe was signed in Washington, marking the start of a major new
initiative which was to earn European astronauts a key role on the US
space shuttle in the early 1980s. And in February 1974 ESRO and the
French space agency CNES finalised an accord on the timetable for the
Ariane development programme, calling for a first launch by the end of the
decade. Ten years after the ill-fated launches of the Europa rocket began,
Europe, nursing its wounded pride, had gone back to the drawing board.

4 Ariane – From Drawing-Board to Launchpad

EARLY ARIANE DEVELOPMENT

The Ariane programme, although eventually to become the most tangible symbol of European co-operation in space, owed its existence from the beginning to the enthusiasm of one country – France. The French government put up 63 per cent of the development costs – estimated in 1973 at 2.47 billion French francs over seven years – and the French space agency CNES became manager of the project.

French state group Aérospatiale was appointed industrial architect, with responsibility for the overall design and assembly of the rocket, and also became prime contractor for all three of the rocket's stages as well as the nose fairing containing the satellite payloads.

Three other French companies also held key prime contracts. Société Européenne de Propulsion (SEP), which had been closely involved in the Europa project, was given responsibility for the propulsion system. Electronics group Matra was assigned responsibility for the equipment bay, which contained the major electronics of the launcher, including the inertial guidance system. And L'Air Liquide was given the specialist task of designing and building fuel tanks for the cryogenic third stage, in which liquid oxygen and hydrogen would be stored at temperatures of minus 180 and minus 250 degrees Celsius respectively prior to and during the launch.

In February 1974 the Ariane programme board met to give the green light for the start of the development phase, and also announced 15 March 1979 as the target date for the first launch. It was an ambitious step, because the plan was to launch the entire rocket with all stages live at the first attempt, without pre-testing individual stages in flight beforehand.

This partly reflected anxiety by France to avoid the political hesitancy which plagued the Europa project by pressing ahead on a tight timetable. But it was also due to confidence by the engineers that Ariane's simpler design would permit a rapid development phase.

Aérospatiale's powerful position as industrial architect for the whole project also gave the Ariane programme much greater cohesion than Europa ever had. While around 50 companies in 11 European countries worked on Ariane, the strong French leadership helped it to avoid the pitfalls inherent in being designed by a wide group of different people.

West Germany financed 20 per cent of the Ariane 1 development, while other countries contributed much smaller amounts. Belgium paid 5 per cent of the total, with the Belgian firm ETCA winning the contract to supply equipment for final checking of the rocket both in France and at the Kourou launch site.

Britain, which had virtually lost interest in building a European rocket after the failure of Europa, put up 2.5 per cent, participating in the project through a bilateral agreement with France rather than directly through ESA.

The Ariane-1 was designed as a three-stage launcher, 47.4 metres high with a lift-off weight of 208 tonnes, of which 90 per cent was fuel, 9 per cent the rocket itself and 1 per cent the payload. To avoid the combined problems of over-complexity and technological experimentation that had spelt disaster for the Europa programme, the policy from the start was to keep Ariane simple and make it work.

> The general architecture of the Ariane vehicle reflects a concern to make maximum use of existing skills and hardware and of technologies that have either been qualified previously in other programmes or represent a minimum development risk

R. Vignelles and P. Rasse, members of the French space agency's Ariane project team, wrote in an article in August 1978.[1]

The rocket's first stage, fuelled by 147 tonnes of UDMH/N204, was to be powered by four Viking motors developed by France's SEP, little different from those built for Europa II's first stage. The engines were designed to produce 245 tonnes of thrust on lift-off, burning for just over two minutes to take the rocket well clear of the South American coast to an altitude of nearly 50 kilometres.

The second stage, also burning for just over two minutes, was to be powered by a single Viking IV engine burning 34 tonnes of UDMH/N204 and developing 72 tonnes of thrust. It was to boost the payload to a height of more than 150 kilometres as it headed out over the Atlantic, and set it on the correct course for the long third-stage burn. The third-stage, standing nine metres high and weighing just over one tonne empty, was to be powered by an HM-7 engine burning a cryogenic mixture of eight tonnes of liquid hydrogen and liquid oxygen. It was the first cryogenic rocket stage built in Europe, and was designed to power the payload for nine and a half minutes as it sped above 200 kilometres in altitude before being released into orbit.

The design of the third stage was a major engineering challenge, since the fuel had to be stored at extremely low temperatures, liquid oxygen at minus 180 degrees Celsius and liquid hydrogen at minus 250 degrees. This required fuel tanks to be made of a light aluminium alloy which was especially reliable at ultra-low temperatures, and to have the two fuel tanks separated by a vacuum. The two tanks were to be individually pressurised during the flight by helium and gaseous hydrogen respectively.

The rocket's course could be corrected during the first stage burn by on-board computers swivelling the Viking engines, while auxiliary jets were built into the second and third stages to prevent launcher rolling in later stages of the flight. The third stage combustion chamber was based on technology developed by Messerschmitt-Boelkow-Blohm of West Germany, technology which was subsequently used in construction of the US space shuttle's main engine.

The new sense of purpose generated by the start of the Ariane programme, after the years of frustration with Europa, helped push development work along at a rapid pace.

The HM-7 engine, one of the most complex parts of the rocket, was given its first test firing by SEP at its Villaroche facility on 27 May 1975. The engine was a prime example of the high-technology Franco-German co-operation which was the driving force for the whole Ariane programme. MBB in Munich was responsible for the injection head and nozzle for the motor's combustion chamber, while SEP handled design and construction of the turbopump and the overall assembly and testing of the engine at its Vernon facility northwest of Paris – the same location to which German rocket scientists had been brought in the aftermath of the Second World War to share the secrets of their technology with the nascent French space industry. Incongruously the Vernon test site, where the roar of rocket motors shattered the calm of the surrounding forest with deafening regularity, was just a couple of kilometres away from a very much more peaceful spot – the lily pond at Giverny made famous by French impressionist painter Claude Monet.

While technical development was getting seriously under way, attention was also being given to the uses to which the new rocket might be put. The aim was to conduct four test flights, starting in June 1979 and continuing into late 1980, to test the rocket's systems prior to charging commercial customers for launching satellites from the fifth flight onwards.

EXPERIMENTAL PAYLOADS

In March 1975 the European Space Agency came up with the suggestion that the second, third and fourth test flights could also be used to carry experimental payloads, to test satellite designs, in-orbit motors or other technological and scientific concepts. Within three months more than 70 ideas had been submitted for experiments for the proposal, which was quickly dubbed Apex (Ariane Passenger Experiments), and ultimately a total of 93 suggestions were put forward.

By mid-1976 many of the proposed experiments had been rejected as too expensive, or unsuitable physically to the constraints of the Ariane payload cone. Five orbiters were being seriously considered for the second Ariane launch (LO2) and a further four for the third launch (LO3), with the fourth launch being reserved for payloads relating to the development of ESA's second generation of applications satellites.

Prime candidates for the second launch included a COS-B satellite, very similar to one launched by a US Delta rocket in August 1975 on a gamma-ray astronomy mission, a GEOS satellite designed to study magnetospheric fields, waves and particles and COSARI, a modification of COS-B to study the magnetosphere and the plasma.

In addition India's experimental telecommunications satellite APPLE and the proposed amateur radio satellite AMSAT were being considered as subsidiary passengers, inserted horizontally next to the main payload.

Main candidates for the LO3 launch in 1980 included an additional Orbital Test Satellite (OTS) platform, separate from the planned operational series of OTS satellites and a prototype version of the Franco-German Symphonie communications satellite series, which had originally been due to fly on the Europa II. Other proposals for launch on LO3 were a Canadian telecommunications platform, an extension of the existing CTS programme to test communication needs in the Arctic north of the country, and an addition to the Meteosat series of European weather satellites.

While the question of how best to use the Ariane test launches was being considered, steady progress was being made both on the launcher and the Kourou space centre in French Guiana, from where it would blast off. Kourou had been the site of the abortive F11 Europa II launch, and therefore already had a viable launch-pad and launch preparation zone which, however, needed uprating to take the larger Ariane.

In March 1976 the 50-metre-high launch tower, weighing 720 tonnes, was raised by 6.5 metres in order to enable it to accommodate the Ariane launcher, which would stand 47 metres high, 15 metres taller than the Europa II. A new section was added underneath the tower and work, although hampered by the Amazon rainy season, continued over the following weeks. Work on new fuel storage tanks was under way while a new road system around the launch site was also being constructed.

DESIGN IMPROVEMENTS

Just as important, a number of improvements to the launch vehicle's propulsion system were being considered, with the aim of slightly raising its payload capacity to 1.6 tonnes into geostationary transfer orbit (GTO) from the previously-envisaged 1.5 tonne limit. The improvements would be achieved by increasing the amount of fuel in the first and second stages, improving the performance of the first- and third-stage engines and reducing the empty weight of the second and third stages. This would enable Ariane to compete better with the US Atlas Centaur rocket for launches of larger telecommunications satellites, such as those being orbited by the Washington-based International Telecommunications Satellite Organisation (Intelsat).

By early 1977 it was possible to announce a further increase in Ariane's intended capacity to 1.7 tonnes from 1.6, partly due to a reassessment of the launcher's capability by the French space agency CNES as a result of new aerodynamic tests. In November 1976 four first-stage Viking engines were successfully test-fired together at SEP's Vernon site, while the same year three tests were also conducted on the newly-completed HM-7 engine for the cryogenic third stage.

In early 1977 companies around Europe began manufacturing the thousands of different parts that would go to make up the first Ariane I, which CNES was planning to assemble by the end of the following year for a June 1979 launch. And in December 1977 ESA was able to state that 'so far no technical problem has emerged which might lead to a slip in schedule'.[2]

GROWING OPTIMISM

Overall, despite a subsequent three-month delay in certification of the first stage, the progress made on the Ariane development meant that by mid-1978 officials were able to look forward with considerable optimism, not only to the following year's test launch but also to the commercial prospects for the rocket beyond that.

Surveys show that the world market for satellites in the period 1981–1990 could be substantial and that Europe is in a position to acquire a significant number of launches, particularly in the applications-satellite field. It would already seem that European missions alone will guarantee two launches per year, a figure that could be raised to four or five by vigourous sales promotion at European level

R. Orye of ESA's Ariane department wrote in 1978.[3]

If anything Orye's forecast was an underestimate, as private sector interest in satellite potential for telecommunications developed fast in the early 1980s.

In April 1978 the ESA council gave the go-ahead for immediate production of five Ariane rockets as a sign of confidence in the programme, while work pressed ahead on both the rocket and the Kourou launch facilities. Engine tests were meeting only mixed success by the end of 1978, however. A test firing of the first stage in early December worked well, showing that it was on target for launch by the following June. But the third stage, the most difficult to perfect because of the cryogenic fuels, was proving more difficult. 'The path towards qualification of the third stage turned out to be rather stony', Peter Creola, Ariane programme board chairman, wrote later.[4]

On 28 November a test of the third stage at the SEP facility at Vernon was aborted after the H8 engine failed to ignite. A malfunctioning safety system caused an explosion on the test stand which damaged the third-stage propulsion bay, forcing a new test schedule to be drawn up, pushing the first launch back to late 1979.

The delay could have been longer, since the damaged third stage was the only one available for final development testing, and taking a replacement through a full test schedule would have delayed the first Ariane launch until the spring of 1980. However, a 1979 launch target date was maintained by a decision to press ahead with the first launch without fully qualifying the third stage – a risk justified by ESA's anxiety to gain in-flight experience of the rest of the rocket's systems as soon as possible.

ESA received a big confidence boost within two weeks of this set-back, however, with the announcement on 7 December that Ariane had won its first major international launch customer – Intelsat. The Washington-based satellite organisation, a true heavyweight among potential commercial clients demanding the highest technical standards, ordered one Ariane rocket for launch by July 1981 for one of its family of satellites, taking an option on a second launch.

> What a breakthrough! A year before its first flight, Ariane had thus been recognised on the international market as a technically and financially viable alternative to the American launch vehicles and the space shuttle

Creola wrote.[5]

With the start of 1979, morale at the European Space Agency started to climb. After the frustrations of the Europa project, the political doubts and the technical hurdles to be overcome, Europe was finally within sight once more of launching its own rocket. Despite the small delay, Ariane was still due to fly before the end of the year and scientists for the first time began to feel that they had the wind behind them.

By February a full-scale white and orange 'propellant mock-up' of the Ariane launch vehicle stood on the pad at Kourou, not the rocket that was to make the first flight, but a simulation to enable engineers to rehearse fuelling and test launch facilities.

> Through the range of tests conducted, the qualification programme of the Guiana space centre has shown up the holes or weak points in the Ariane launch system at all the various levels of operations, and enabled the necessary corrective actions to be taken ... it is a perfectly operational space centre which is now ready to receive the first rocket

M.-A. Hauzeur, of the ESA's Ariane team in Kourou, wrote in the late summer of 1979.[6]

And on 26 July the ESA council took two important decisions for the future of the Ariane programme. The first was to order components for five further Ariane rockets, taking the total number of rockets in production to 11 from six. And it signalled the start of the next stage of Ariane development, directing its scientists and engineers to upgrade the Ariane design to produce a rocket capable of launching payloads weighing 2.35 tonnes into geostationary transfer orbit (GTO). This would represent an improvement of nearly 20 per cent on the Ariane I performance.

LAUNCH PREPARATIONS AND COUNTDOWN

On 12 September, three months before the planned launch date, five white containers holding the three fully-assembled rocket stages and the two half nose-cones were loaded onto a barge on the Seine at the Aérospatiale assembly plant at Les Mureaux, west of Paris. Three days later the freighter *Carimare*, with the containers on board, slipped her berth at Le Havre, swung west out of the harbour and steamed towards the Atlantic, bound for Cayenne, French Guiana.

In late September the ship arrived at Cayenne, where the containers were loaded onto trucks and driven the 60 kilometres up the coast to the space centre, past groups of curious villagers scattered along the thin strip of road sandwiched between the jungle and the ocean. The launch campaign began officially on 1 October, with a target launch date of 15 December. And following a series of pre-flight tests, with the green light given for launch, fuelling of the rocket started mid-morning on Friday 14 December.

A tropical rainstorm was lashing the coast of French Guiana as launch day dawned. At 05.40 the protective tower started to roll back from the rocket, leaving it silhouetted against the sky with the steaming jungle waking up all around. The launch had been set for 11.00 local time (14.00 GMT), but an early morning conference on the weather led to a decision to postpone it by half an hour to give the rainclouds a chance to clear. As the countdown continued and the moment of lift-off approached, tension mounted palpably both at the subterranean launch control centre a few hundred metres from the pad, and at the 'Jupiter' control centre 12 kilometres down the coast, packed with flight controllers, dignitaries and journalists. Crowds lined the beaches down the coast. 'Saturday morning fever', one French newspaper had proclaimed, capturing the feeling that a great deal more was riding on the launch than the success of an individual rocket, and that once again it was the credibility of the whole European space programme that was on trial. In Paris French President Valéry Giscard d'Estaing took time off to watch the launch preparations on a satellite link.

At 1124 the six-minute automatic countdown sequence to blast-off started, with two computers, one controlling the electrical systems and one the propellant and fluid systems, on a synchronised count.

With a minute to go the rocket switched onto its internal power systems, just seconds before ignition the cryogenic arms linking the launch tower to the rocket swung away, and at zero plus 0.6 seconds the four Viking engines powering the first stage ignited. Within three seconds the rocket was due to rise from the pad, clearing the surrounding forest on a plume of flame. But nothing happened. A cloud of smoke from the engine burn rose above the pad, but at zero plus eight seconds the engines shut down. Disbelief swept the control room.

In fact the sequence of events leading to the aborted launch was frustratingly simple, and rapidly analysed. All four engines had fired correctly, but a malfunction in an on-board monitoring device meant that misleading information was fed to the computers, indicating that engine A was functioning at below normal pressure. The computer, armed with this false information, automatically shut down the launch.

Technicians at the Guiana space centre had already rehearsed the procedure for launch abort, and immediately set about draining the volatile liquid oxygen and hydrogen from the third-stage tanks. Problems in maintaining the correct pressure dragged out the process, with the hydrogen tank taking nearly eight hours to empty and the oxygen tank 15 hours. The first- and second-stage tanks were also drained to prevent corrosion, and the launch was put back more than a week to allow time for technicians to arrive from France to inspect the rocket for damage.

SECOND LAUNCH ATTEMPT

On 17 December technicians wearing breathing apparatus conducted an internal inspection of the Viking engines to check that they had not been damaged by the searing heat during the ignition and that the rocket was safe to fly. The following day ESA technicians were sufficiently confident to reschedule the next launch attempt for within five or six days, despite some lingering doubts about the engines, and after a final review later that week launch was set for 11.00 local time on 23 December.

Two days later fuelling of the first and second stages started again, and on the morning of 23 December liquid oxygen and hydrogen was once more pumped into the third-stage tanks. With the launch put back to 12.00, the countdown progressed smoothly until just 52 seconds before lift-off, when a computer stopped the automatic sequence, again due to a problem associated with monitoring the rocket rather than the actual state of the rocket itself. The interruption in the count led to a problem with the pressurisation of the third-stage helium sphere, which was designed to maintain the low-temperature fuel in a stable condition, and while engineers were trying to tackle this the weather deteriorated. As a result, with a tropical downpour drenching the launch site, the decision was taken in early afternoon to postpone the launch attempt until the following day.

The morning of Christmas eve dawned bright and clear, with technicians aware that time was running out if the rocket sitting on the launchpad was ever to get airborne. Some hydrogen fuel had been consumed during the fuelling and defuelling process for the two attempted launches, with the result that failure to launch on 24 December would have meant ordering new supplies of hydrogen. It was also unclear to what extent the rocket's fuel tanks were becoming corroded, and failure to launch by New Year could force ESA to scrap the rocket on the pad and send for a replacement.

The countdown progressed into the early afternoon until, with barely two minutes to go, the computers again decreed a halt. This time the problem lay with a pressure valve on the third-stage hydrogen tank, which had closed too slowly. An attempt to restart the count was initially foiled by one of the ground computers, which refused to transmit the order – so technicians were forced to improvise rapidly to override the faulty circuit. Then a pressure drop in the third stage helium sphere caused concern, as it had the day before, but this time the problem was overcome.

The count restarted at the six-minute mark, passed through the automatic sequence, and at 14.14 and 28 seconds local time (17.14 GMT) the four first stage engines roared into life. Orange smoke billowed out over the surrounding jungle as, 3.4 seconds later, the rocket rose slowly from the pad, ponderously at first, then gathering speed. Seven seconds later it was clear of the tower, and at the 24-second mark it rolled out of its vertical climb to arch out over the ocean, applauded by spectators lining the beach down the coast from Kourou to Cayenne.

Two minutes 25 seconds into the flight the first stage burn ended precisely on schedule, the rocket climbing to a height of 47 kilometres and hitting a speed of more than 6500 kilometres per hour. It was still just visible from the launch site as the first stage dropped away into the ocean and the second stage Viking 4 engine ignited. As the rocket continued on its fiery path it was being tracked, initially by the Kourou telemetric station, then the tracking station at Natal, Brazil, and, as it headed towards the west coast of Africa, the US-operated tracking station on Ascension Island.

Nearly five minutes into the flight, with the rocket now flying at more than 12 000 kilometres an hour, came the critical second-stage separation and third-stage ignition. The most advanced piece of rocket technology ever built in Europe was about to be put to its first flight test. More than a few anxious minds flashed back to that day in November 1971 when the Europa II rocket had plunged into the Atlantic, shattering European space aspirations for nearly a decade. Would Ariane be a repeat performance?

The third stage fired perfectly, accelerating the rocket up through a speed of more than 20 000 kilometres an hour as it sped towards Africa, climbing towards its peak altitude of 206 kilometres. The flight path, although slightly higher than planned, was near-perfect, and everything seemed to be going smoothly until around seven minutes into the third-stage burn, when the electronic trajectograph at the Kourou control centre started to show erratic movements of the rocket. There was a flurry of concern for the safety of the launcher, until the explanation came through on the public broadcast system. Differences in measurement by the tracking stations at Natal and on Ascension Island had produced jagged lines on the trajectograph, but the rocket itself was in good shape.

The rest of the flight went perfectly, with the payload bay opening out to inject the experimental CAT satellite into an accurate elliptical orbit. The sole function of the satellite, which had a life of less than three days, was to measure its own orbit and verify the success of the flight.

As the mission control officer announced the separation of the satellite a great cheer went up in the Ariane control room. Within a couple of hours champagne corks were popping at a party in the Ariane assembly hangar as the frustrations of the past decade were swept to one side. As Peter Creola, chairman of the Ariane programme board, wrote later

> Excitement, hugs and tears – looking back, this now seems oversentimental, but at the time it was certainly the most moving and historic moment in twenty years of European space cooperation . . . After all the mistakes and all the crises, Europe, joined together in ESA, now had its own launch vehicle, and hence the essential means to conduct its own space policy.[7]

The Ariane launch success came just a week before the start of the new decade and gave the European space programme the boost it needed to approach the 1980s with a confidence that until then had been in short supply.

5 Early Scientific Satellites

As evening fell over NASA's Western Test Range in California on 29 May 1967, an American Scout rocket streaked into the Pacific sky to make what those watching on the ground hoped would be a piece of space history. Aboard was the ESRO-II satellite, the first pan-European research satellite and the fruit of several years of design work and scientific debate. It was supposed to study cosmic rays, solar X-rays and the composition of the Earth's radiation belts.

But barely three minutes after launch a malfunction in the rocket's third stage, followed by the failure of the fourth stage to ignite, destroyed the launcher and its pioneering payload. Instead of celebrating a landmark in space co-operation, scientists watched their tracking screens helplessly as the rocket disintegrated over the ocean.

It was an inauspicious start to the European Space Research Organisation's ambitions to become a serious competitor in the space race. But it was not the first blow that it had suffered.

Just over 18 months earlier, on the night of 14 October 1966, a fire had swept through a temporary pre-fabricated building at the European Space Technology Centre at Noordwijk, the Netherlands, destroying offices and facilities being used by ESRO's satellite department. Apart from a mass of documents that were lost, the fire also destroyed a simulator used in research on large satellites, damaged a computer in the applied mathematics division and created havoc in the environmental testing division. The ESRO annual report for 1966 was relatively sanguine about the effects of the fire, commenting that

> as far as can be foreseen, none of the scheduled launch dates of the satellites on the current programmes are likely to be affected.[1]

But it did predict delays of up to a year in some work on materials testing, due to loss of equipment.

At the time ESRO was pressing ahead with work on several fronts. Not only was the ESRO II satellite project in preparation, and its sister satellite ESRO I, which was designed to study the auroral zones around the Earth, where phenomena such as the spectacular Northern Lights are produced as a result of geomagnetic and solar activity; but work was also nearing completion on the HEOS-1 satellite (Highly Eccentric Orbiting Satellite), whose orbit was to take it nearly a quarter of a million kilometres from the Earth to research into the interplanetary magnetic fields and the behaviour of the solar wind.

In addition preparatory work was progressing on an ambitious twin satellite project, TD-1 and TD-2, which were to break new ground in ultra-violet, X- and gamma-ray astronomy. There was no shortage of enthusiasm, nor inventiveness, for new scientific projects in the organisation, and scientists were already beginning to look beyond the satellites in development to the design of a much more ambitious Large Astronomical Satellite which they hoped would become Europe's first large-scale orbiting observatory.

ADMINISTRATIVE AND FINANCIAL PROBLEMS

ESRO's scientific ambitions were beginning to run ahead of both its management capabilities and its financial resources, however, contributing to a tension within the organisation that was never far beneath the surface during the mid-1960s.

> The most dangerous clouds over ESRO's future were undue expectations of what European industry could achieve under ESRO's guidance within the financial limits set and within a relatively few years. Neither the size of the management task nor the cost factors were understood

Sir Hermann Bondi, who was ESRO's director-general from 1967 until 1971, recalled later.[2]

One aspect of the problem was that unrealistically low cost forecasts for satellite programmes had been made, while a second was the centralisation of decision-making at ESRO, with the supervisory Council involving itself too much in the day-to-day running of the organisation. The council was split by disagreements over budget plans from 1967 onwards, forcing it to take a rather piecemeal approach to planning since it was unable to reach the unanimous agreement required under the ESRO convention to approve the three-year budget for 1967/69.

The council tackled the problems by appointing a group of experts chaired by a Dutch official, J. H. Bannier, to propose reforms of the organisation. The group completed its report in March 1967, and submitted it to ESRO Director-General Auger.

> The essential conclusion was that the organisation suffered from excessive centralisation of powers, and the fundamental recommendations contained in the report suggested that the council and its committees should delegate a large part of their power to the director-general

an ESRO report said.[3]

The Bannier report proposed a less centralised structure for the ESRO directorate, and a simpler, more flexible management style, which it believed would bring decision-making closer to the teams in the field.

> If the directorate cannot be situated wholly at the place where all or most

of the executive work is done, we prefer it to be itself geographically spread rather than to maintain a separate headquarters to which reference has to be made for all or most decisions

the Bannier report said.[4]

The report was rapidly implemented, with the management split into a directorate based in Paris responsible for strategy and planning and two executive directorates, one based at the ESTEC technical research centre at Noordwijk and the other at the satellite control centre at Darmstadt. Five directors were named by the end of 1967, enabling ESRO to enter 1968 with a reformed structure and a more self-confident image. Auger, who had been director-general since March 1964, bowed out in October 1967 and handed over to Sir Hermann Bondi, who was to guide ESRO through its early period as a satellite operator.

The satellite programmes that ESRO was working on in the mid-1960s were diverse, and relatively ambitious given the lack of European experience in the field of operational satellite systems. In the light of the failure of its sister organisation ELDO to make significant headway during the period in its work on developing a credible European satellite launch vehicle, ESRO scientists were able to take considerable pride in their work, even if it was subject to frustrating delays due to political difficulties or occasionally wavering financial support.

Scientists looking back regard 1968 as something of a breakthrough year for the European space programme, for it was in that year that the continent orbited its first major scientific satellites and started to map out a course for itself that did not simply mimic aspects of what the US and Soviet space programmes had already achieved. Four years after the ELDO and ESRO conventions formally came into force in 1964, tangible successes were at last being achieved.

THE ESRO AND HEOS SATELLITE PROGRAMMES

The failed launch in May 1967 of the first ESRO II satellite not only marked a setback for the scientists who had experiments riding on board the pioneering satellite, but was also a blow to the credibility of the ESRO organisation as a whole. Therefore engineers intensified their efforts to prepare an identical sister satellite, the ESRO II F2, otherwise known as Iris, for launch as soon as possible.

The tiny satellite, which was only 86 centimetres long and 78 centimetres in diameter, was to carry seven experiments, studying the way that the immediate environment around the Earth is affected by the sun. One set of experiments was to study X-rays, another the way radiation becomes trapped in the very high energy Van Allen belts around the planet and a third the electronic composition of cosmic rays. Five of the experiments were British, one Dutch and one French.

While investigation of the Earth's magnetosphere, that part of space under the magnetic influence of the Earth, was not a subject calculated to command much public attention, the satellite was to pioneer research into a little-understood area, shedding light on phenomena ranging from aurorae to radio static.

> Our understanding of the magnetosphere and the Sun is probably very similar to Columbus's understanding of the Atlantic Ocean and America after his historic voyage

D. E. Page, ESTEC's space science head, wrote.[5]

Work on preparing the second flight model of ESRO II took around four months and was completed in December 1967. After tests at Noordwijk it was shipped to the Western Test Range in California where, on 17 May, it was successfully launched on a NASA Scout rocket of part-Italian construction. The satellite was placed very close to its optimum orbit, circling the Earth every 99 minutes with a 1090 kilometre apogee (maximum altitude) and 330 kilometre perogee (minimum altitude). Scientists flying back to the satellite control centre at Darmstadt to activate the experiments were elated. After the years of planning, hard work and missed opportunities, Europe as a community of nations finally had its own, albeit modest, satellite in orbit.

Once the first success had been achieved, the second was not long in following. On 3 October the same year a second satellite was successfully orbited, again by a Scout rocket. This vehicle, ESRO-1A, had a different mission from its sister vehicle, being designed instead to study the ionosphere and related auroral and polar cap phenomena. Auroral phenomena, known as aurora borealis or Northern Lights in the northern hemisphere and aurora australis in the southern, are caused by the collision of highly charged particles in the solar wind with gases in the upper atmosphere.

The phenomena occur between around 65 and 70 degrees north and south of the equator, towards the polar regions, and the task of ESRO-1A was to measure how auroral phenomena moved as a result of changes in geomagnetic and solar stimuli. The satellite carried eight experiments, four supplied by British research institutes or universities and four by Scandinavian.

> In the years just preceding spacecraft flights it was suspected that solar particles of some sort arrived over the polar cap. ... however, major advances were made by the combined efforts of experimenters on HEOS-1, ESRO-II and ESRO I

ESTEC's head of space science commented.[6]

Apart from the success of individual experiments, space officials welcomed the willingness of scientists working in different countries to exchange data, helping to build up a mass of knowledge at an early stage in ESRO's scientific programme.[7]

5 December 1968 saw the launch on a US Thor-Delta rocket of the third joint European satellite, HEOS-1, which was put into an eccentric five-day orbit taking it nearly a quarter of a million miles from Earth at its farthest point. Its role was to study the Earth's magnetic field, cosmic rays and interplanetary physics. Its ability to measure the direction and energy of particles heading for the Earth, whose arrival was being studied by the ESRO satellites already in orbit, meant results from the three satellites could be collated to give a picture of the magnetosphere.

In fact HEOS-1 considerably out-performed expectations, staying in orbit for nearly seven years.

> Its full payload gave excellent scientific data for about 16 months, exceeding the design lifetime of the satellite by four months. ... with the magnetic field experiment continuing to work until re-entry, thus providing an unprecedented record of the interplanetary magnetic field for almost seven of the eleven years of a solar cycle

a project scientist wrote later.[8]

ESRO scientists, after the frustrations of the early years of the organisation, now had three satellites finally in orbit and transmitting data. As Professor Henk van de Hulst, President of the ESRO council, wrote in its 1968 annual report,

> In 1968 the long years of preparatory work have started to bear fruit and ESRO has become a totally operational organisation. By the end of the year we had three satellites in orbit carrying 22 experiments which were all working well and transmitting valuable information. These satellites are the most complex ever conceived, assembled and operated by Europe. This achievement, by an organisation which has been in existence for less than five years, can legitimately inspire satisfaction and pride in all those who have contributed to it.[9]

TD-1 AND TD-2 ASTRONOMICAL SATELLITES

However, at the same time as ESRO was congratulating itself on finally getting some experiments into orbit, it was facing the familiar problem of cost overruns on another, rather more ambitious project. TD-1 and TD-2 had been designed as twin astronomical satellites, carrying around four times the weight of experiments as the ESRO-I and ESRO-II vehicles, with TD-1 designed to study radiation from the sun and other stars and

TD-2 the interaction between solar activity and the Earth's ionosphere. Launch of both satellites was originally planned for 1970, but by early 1968 it was becoming clear that the project was going to be considerably more expensive than originally envisaged, even using essentially the same design for the two satellites. By 1968 ESRO's budget had risen to Ffr. 264 million, eight times its 1964 start-up budget of Ffr. 34 million, with spending specifically on the satellite programme accounting for just over Ffr. 100 million of the total. Most of the rest went on personnel costs, administration and the upkeep of facilities, with a relatively modest Ffr. 23 million being spent on sounding rockets.

To hold down costs, it was decided to simplify the TD-1 mission and scrap TD-2, holding over some of the experiments planned for TD-2 for a later satellite. In the meantime two more satellites were launched.

ESRO-1B was sent into space in October 1969 to study the same auroral phenomena as ESRO-1A but from a lower orbit. Then in January 1972 HEOS-2, the sister satellite to HEOS-1, was launched on a mission to study the polar magnetosphere and interplanetary space.

12 March 1972 brought the long-awaited launch of TD-1, Europe's first attempt to place telescopes outside the Earth's atmosphere to scan the stars. The telescopes, by far the most advanced ever flown at that time, concentrated on ultra-violet radiation. Since most of this is absorbed by the Earth's atmosphere, ultra-violet astronomy is not practical from Earth, so the chance to place telescopes in orbit opened up a whole new area of research.

TD-1 worked well from the start and scientists were happy with the data, but barely two months after launch the project, in which millions of dollars and years of work had been invested, was facing disaster. While the satellite was harvesting huge quantities of data, it could only transmit a fraction back to Earth because both its tape recorders had broken down.

The tape recorders were vital, because data could only be captured on Earth when the satellite was passing over a receiving station. All data amassed while the satellite was out of range, as it was for over three-quarters of its orbit, had to be recorded on board for later transmission if it was not to be lost.

The problem was that only seven stations had been planned to receive data from the satellite along its orbital route. An appeal for help to NASA and CNES doubled the network of ground stations, but three-quarters of the data was still being lost. So in June, three months after the launch, an ambitious rescue plan was put into action to involve more existing ground stations and set up, in the space of a few weeks, a global network of mobile tracking stations to save the mission. By 1 August more than 30 stations were functioning, boosting the amount of data being received from 15 per cent of the satellite's output to more than 60 per cent. Hastily-assembled mobile stations were flown out around the world and, once operational, both the equipment and the teams operating it had to cope with a variety of extreme climatic conditions ranging from the heat and humidity of Singapore to hurricanes on Fiji and the icy wastes of Antarctica.

One mobile station was even set up on a Dutch banana boat, the *Candide*, in the Pacific, in an area where it was impossible to establish a land station. As ESRO director-general Alexander Hocker recalled later

> M. S. Candide was stationed in the roaring forties, where throughout the mission the wind averaged force seven, and at times reached force 11. Part of the antenna was carried away, and the ship was continually rolling and pitching, making both operations and maintenance extremely difficult. The cramped accommodation, vibration and limited food and water added to the rigours of the voyage.[10]

The operation continued until the end of October, when the satellite went into eclipse. But in February the following year it was reactivated, with the network of ground stations this time recovering close to 70 per cent of the data. The ESRO annual report of 1972 paid tribute to 'an unprecedented rescue operation ... a truly international effort'.[11]

The episode highlighted the enthusiasm and inventiveness of the agency's staff, determined to salvage a project heading for disaster.

Although TD-1's sister satellite, TD-2, never flew, some experiments which had been designed for it were launched aboard the ESRO-IV satellite in November 1972 to study the relationship between the neutral atmosphere and the ionosphere. So by the end of 1972, although the early satellites had ceased to function, ESRO had four operational scientific satellites in orbit – the two HEOS satellites, TD-1 and ESRO-IV.

COS-B AND COSMIC RADIATION

While scientists continued to work on developing projects for future missions, Europe's failure to match its success in building satellites with any significant advances in rocketry was leading to a major shake-up in the organisation of the European space effort. Scientists, engineers and politicians all agreed that Europe's way forward lay in combining the functions of ESRO and ELDO in a single organisation for rockets and satellites. And so it was that when Europe's next satellite was placed in orbit, in August 1975, it was under the auspices of the new European Space Agency.

Apart from the political upheaval that formed a backcloth to its development phase, the satellite, COS-B, had been particularly difficult to assemble and test. This was partly because of its technical complexity, since six different scientific institutes were working on the project, and partly because of intermittent funding problems. Each institute was funding its role in COS-B separately, and when funding for one of the groups unexpectedly dried up, ESA had to make urgent arrangements for ESTEC's space science department to take over responsibility for the experiment. 'COS-B has provided more headaches than any other previous satellite', ESA's first director-general Roy Gibson wrote following the launch, adding that he was determined ESA should provide central financing for the hardware on all future observatory satellites.[12]

COS-B's scientific purpose was to gather data on gamma rays in order to shed light on the origin of cosmic radiation. Cosmic rays are produced by events such as exploding stars, or supernovae, and solar flares, but their point of origin is difficult to determine because their paths are deflected by electric and magnetic fields in space. Because gamma rays are produced by cosmic radiation, but cannot be deflected since they are uncharged, the direction from which they arrive at Earth indicates the origin of cosmic rays.

The multi-national co-operation that went into the launch of COS-B was typical of the growing complexity of European space operations, and illustrated the degree of interdependence which ESA and its forerunner organisations had fostered throughout the continent. Construction of the satellite itself was the responsibility of Messerschmitt-Boelkow-Blohm of West Germany, but a whole range of companies were involved with its production, including Aérospatiale of France, BAC and Ferranti of Britain, Selenia and Laben of Italy, ETCA of Belgium, CASA of Spain, Fokker of the Netherlands and Terma Electronik-Centralen of Denmark.

The scientific payload was the overall responsibility of ESTEC at Noordwijk, the Netherlands, but individual experiments were prepared by the Saclay nuclear study centre in France, Laben and the Universities of Milan and Palermo in Italy and the Max Planck Institute for physics and astrophysics in West Germany. Satellite check-out equipment was provided by ESTEC, Selenia of Italy and Marconi and Honeywell in Britain, and the satellite was launched from a NASA pad in California on a Thor Delta rocket.

That there was nothing particularly unusual in the complexity of the operation is evidence of the high level of organisation that was required by the mid-1970s to get ESA projects into orbit.

COS-B had a useful life of more than six years and by the time it stopped transmitting data, in April 1982, ESA had launched four more scientific satellites, was starting to put in place telecommunications and weather satellite systems and had its own operational rocket programme. There was a sense in the agency that after the frustrations of the 1960s and the disappointments and upheaval of the early 1970s, the European space programme was at last on the move.

Nevertheless, there was a 20-month gap between the launch of COS-B in August 1975 and the next launch, of GEOS-1, on 20 April 1977. And the launch, far from being smooth, was the most problematic since the breakdown of the TD-1 tape-recorders five years earlier.

THE GEOS PROJECT

GEOS-1 had been designed as ESA's first scientific satellite to go into geostationary orbit, and as a result had been selected as a reference spacecraft for the International Magnetospheric Study, a global programme using ground observatories, balloons and sounding rockets to study the magnetosphere – the part of space where the Earth's magnetic field interferes with the ionised particles in the solar wind, stretching out around 60 000 kilometres on the sunward side of the planet and much further on the far side.

So it was particularly frustrating when the Thor-Delta rocket which launched GEOS-1 failed to put it into a transfer orbit from which geostationary orbit could be achieved.

> The first indication of failure was received in the Mission Director Centre at Cape Canaveral about 25 minutes after lift-off. . . . some time later a NASA orbit prediction gave an apogee height of 12 000 kilometres – only one-third of that required for injection into a geosynchronous orbit

a post-launch ESA report said.[13]

A clamp band binding the second and third stages of the rocket had opened early, so that although the third stage ignited it lacked spin, preventing it from putting the satellite on its proper trajectory.

Project managers at ESOC in Darmstadt succeeded in stabilising the satellite in a holding orbit and five days later fired it into an elliptical orbit looping as far out as 38 300 kilometres from the Earth, compared with its intended circular geostationary orbit with an altitude of just over 36 000 kilometres.

The mission was partially saved, although the satellite was able to supply much less data than intended because most of its orbit was still useless. So ESA rapidly decided to go ahead with a launch the following year of a back-up satellite, GEOS-2, with the same experiments on board. This satellite was launched in July 1978 and successfully functioned until late 1983, when it was fired into a marginally higher parking orbit. Even there it continued to perform a back-up role for studies on the magnetosphere being carried out by more technically advanced satellites.

6 Early Communications and Weather Satellites

In parallel with development of a comprehensive scientific satellite programme, European countries were also working on a separate area which was to become within a few years the leading growth area of the space industry – telecommunications satellites.

It was France which in the mid-1960s had pushed its sceptical European partners into taking a serious look at the possibilities offered by applications satellites. At the time few people foresaw the potential for the development of totally new telecommunications services on satellites, regarding them rather as vehicles for supplying existing services, primarily telephonic, in a perhaps cheaper or more efficient form.

The first serious discussion of the idea of setting up a pan-European satellite telecommunications network came in 1964 at a meeting in Bonn of the newly-formed European Conference on Telecommunications by Satellite (CETS). Just two years earlier the world's first telecommunications satellite, Telstar, had been placed in orbit, beaming the first live television pictures across the Atlantic and, despite the primitive nature of the black and white images, vividly demonstrating the possibilities of the technology.

In 1965 Intelsat, the Washington-based international organisation formed to provide a global communications system, launched its first satellite, Early Bird, into geostationary orbit, placing the concept of telecommunications satellites on an operational footing. But it still took a couple of years for Europe to stir itself to action.

The delay was partly because the continent, having established one organisation to develop launchers and another to build scientific satellites, had not created a body with any responsibility for applications satellites. It was at France's insistence that the design specifications for the embryonic Europa 1 rocket were changed in April 1965 to enable it to launch telecommunications as well as scientific satellites.

Then in 1967 European countries started to explore the possibilities of building their own telecommunications satellites, with two separate initiatives running in parallel. France and West Germany began work on an experimental telecommunications satellite programme called Symphonie, while on a broader level the European Space Conference, grouping ministers from the participating countries, supervised a separate project involving ESRO, ELDO, the European Broadcasting Union (EBU) and the European Conference of Postal and Telecommunications Administrations (CEPT).

The Franco-German project came to fruition more quickly, with the first

Symphonie satellite being launched in 1974 and the second the following year. Meanwhile, political endorsement for the pan-European initiative came at a meeting in Brussels in 1970 of the European Conference of PTT ministers, and in 1972 nine countries gave their formal approval to press ahead with a telecommunications satellite programme, involving the construction of an experimental satellite OTS (Orbital Test Satellite), to be followed by two operational ECS (European Communications Satellite) vehicles later.

THE ORBITAL TEST SATELLITE (OTS)

The OTS satellite, which was to form the basis for a whole generation of communications satellites, was a major technological challenge. It was hexagonal, with a lift-off mass of 865 kilos, of which around half was accounted for by the apogee motor and fuel needed to boost it into geo-stationary orbit. Its purpose was to demonstrate the technological feasibility for the European PTTs to offer existing services by satellite, but also to encourage the growth of a range of new data transmission services, some through the PTTs and some to private customers equipped with their own reception dishes. These dishes, although much smaller than the large PTT antennae, would still be three metres or more in diameter, very cumbersome compared with the mobile dishes to be developed by the end of the 1980s.

The satellite was designed and built by the MESH consortium, which won the contract in November 1973. The consortium was led by Hawker Siddeley Dynamics of Britain, with five principal co-contractors spread around Europe. These were Matra of France, ERNO and AEG-Telefunken of West Germany, Saab of Sweden and Selenia of Italy, with a further 40 sub-contractors. Matra was responsible for the final assembly and testing of the satellite, as well as the electrical ground-support equipment and the attitude and orbit control subsystem, while Hawker Siddeley took control of the satellite's power system, electrical distribution and mechanical ground-support equipment. The orbiter's structure was to be built by Aeritalia, another MESH member, under sub-contract from ERNO, which also had responsibility for the reaction control system, while Saab-Scania designed and built the telemetry, tracking and command systems. Selenia supplied antennae and AEG-Telefunken repeaters.

The satellite was to be tested in a variety of ways, with European Space Agency (ESA) scientists monitoring the actual performance of the orbiter and a range of potential users carrying out experiments designed to evaluate the capabilities of systems developed for it. The European Centre for Nuclear Research (CERN), for example, proposed a relatively simple data transmission experiment, linking its scientists around Europe to computers in other centres, while a Stockholm newspaper took a data channel on the satellite to experiment with printing editions in various other parts of Sweden, a fairly revolutionary concept at the time. Limited experiments with transmission of television pictures were also proposed, although this was not the main purpose of the satellite.

The first launch attempt, in September 1977, was a complete disaster, with the Delta rocket carrying OTS 1 exploding just seconds after lift-off. It took eight months to prepare the back-up model, OTS-2, for launch, but on 11 May 1978 another Delta placed it into a perfect geostationary orbit over the equator at 10 degrees East, high above Gabon and with an ideal vantage point for pan-European cover. The arrival of the satellite in its planned orbit, almost ten years to the day after Europe had orbited its first scientific satellite, marked the graduation of the European space programme from a challenging, but arguably esoteric, exercise of interest mainly to the scientific community to a programme aimed at developing telecommunications systems for a mass market not only in Europe but also further afield.

As E. S. Mallet, chairman of ESA's joint communications board, oberved once the satellite had been placed in orbit:

> The launch of OTS is a step towards a goal of a European regional communications satellite system ... although this is a valuable achievement in itself, the ultimate goal is a share of the world market in regional satellite systems. Although it can be said now that European firms have a high technical competence in the telecommunications satellite field, they have yet to show their commercial capability. ESA must continue to maintain a level of investment which will enable member countries to achieve this aim, one that is unlikely to be won by any member country on its own.[1]

The OTS satellite was originally designed to stay in orbit for three years, testing technology for a family of operational satellites to be launched subsequently. In fact it continued to function for more than a decade.

Four families of satellites were subsequently built and launched that were designed around the basic OTS model. The immediate product was the series of European Communications Satellites (ECS), which were to provide telecommunication services across Western Europe from the early 1980s. Related to this was the decision to develop a maritime version of the ECS series, called MARECS, to serve the need of oceanic communications. The Skynet series of military satellites for the British forces, and the identical NATO orbiters for The Atlantic Alliance, were also based on the OTS design, as were the Télécom satellites built by France to serve both its civilian and military communications needs. By the late 1980s a total of 16 telecommunications satellites had been built on the OTS principle, and most of them successfully placed in orbit.

While the OTS satellites were still being prepared for launch, the question arose of who was going to maintain and operate the satellites, lease out space on them and take responsibility for the whole administrative side of the project. In 1977 the 17 Western European members of the European Conference of Postal and Telecommunications Administrations (CEPT) signed an agreement setting up an interim version of an organisation to be known as Eutelsat. Satellites provided an opportunity, but also a threat, to the existing telecommunications operators, state monopolies which were used to handling land traffic and feared that the new technology becoming available through the initiative of the European Space Agency could slip out of their control.

> Eutelsat was born because of the need for the PTT administrations to protect their exclusive right of providing telecommunications throughout Europe ... so the PTTs, as a defensive move, decided to build the organisation

Andrea Caruso, director-general of Eutelsat, recalled.[2]

France, Switzerland and Italy were the countries showing the most immediate enthusiasm for the new technology, and were the driving force in setting up the Eutelsat organisation. But Britain and West Germany, for different reasons, had doubts about the benefits of satellites. West Germany stood to lose commercially, at least in the short term, since due to its geographical position at the centre of Europe it was able to levy transit fees on international telecommunications traffic passing through its land lines, business which satellites threatened to take away. In Britain, Caruso said, there was disagreement between the government, which could see the potential of the OTS programme and was financing ECS development within the European Space Agency, and the British Post Office, which remained to be convinced that satellites could significantly contribute to improving communications in Europe.

> The UK hostility – at Post Office level – was due to the fact that they did not believe in the need for space communications in Europe ... they did not believe that there was going to be a profitable investment in this project

he said.[3]

Eutelsat operated as an interim organisation throughout the late 1970s and early 1980s, only becoming a formally-ratified organisation in September 1985 when the Eutelsat convention came into force. Headquartered in Paris, its membership had grown to 26 national PTT authorities by the end of 1980s from the original 17 signatories, with four satellites successfully launched.

THE METEOSAT PROGRAMME

Apart from telecommunications satellites, the other major area of satellite applications being investigated in Europe during the mid-1970s was that of meteorology.

> In the middle sixties we witnessed a move from pure scientific research towards the development of space applications. Similarly, in the coming years, the move will be from research and development towards operational uses. This is particularly the case for telecommunications and meteorology

A. Lebeau, ESA's director of planning and future programmes, wrote in 1976.[4]

The Meteosat programme, Europe's first venture into satellite-based weather forecasting and Earth observation, started as a French programme in the late 1960s, but was taken under the wing of the embryonic European Space Agency in 1972, expanding to eight the number of countries involved in the project – France, West Germany, Britain, Italy, Switzerland, Sweden, Denmark and Belgium. Scientists responsible for the project were based in Toulouse, France, with about two-thirds of the Project staff coming from ESA and one-third from the French space agency CNES.

Meteosat was to form the European contribution to a world-wide weather satellite system comprising five satellites. It would be stationed over the Greenwich meridian to cover Europe and Africa, as well as the Mediterranean, Middle East and eastern Atlantic Ocean. The Soviet Union was to provide a satellite covering Asia and the Indian Ocean, Japan a satellite covering Australia, the Far East and the western parts of the Pacific, and the US two satellites, GOES East and GOES West, monitoring North and South America, the western Atlantic and eastern Pacific Oceans.

Early meteorological satellites launched by the US during the 1960s were placed into low orbits, circling the Earth in less than two hours but only able to monitor weather over the same point once every 12 hours. In 1974 the US launched the first geo-stationary weather satellite, SMS-1, which transmitted pictures of the continental US back to Earth every half hour, permitting meteorologists to put together much more dramatic and informative sequenced images of cloud movements.

Design work on Meteosat started in 1974, with the project being led by French state group Aérospatiale, and during the following three years tests were carried out on engineering models and a prototype, Meteosat-1, was constructed. Its launch on board a US Delta rocket was subject to several delays due to interference signals being found in the same frequency band as the rocket's telecommand in-flight destruct signal, interference which was ultimately traced to a leak from a power amplifier on a US ship.

This Meteosat launch was the first one after the failure of the OTS-1 launch in September 1977 and ... neither NASA nor the (European Space) Agency was ready to take the slightest risk

P. Goldsmith, director of ESA's Earth observation and microgravity programme, said later.[5]

It was successfully launched on a Delta rocket on 23 November 1977 and was manoeuvred into its correct parking orbit by early December. Equipped with a scanning radiometer, its job was to transmit high-resolution pictures of its section of the globe every half an hour, with a resolution of 2.5 kilometres in the visible spectrum and five kilometres in the infra-red. The pictures were transmitted to receivers at Odenwald, West Germany, before being processed at ESOC in Darmstadt and retransmitted to end users back through the Meteosat orbiter.

Apart from forecasting weather over Europe and Africa, Meteosat's role in the global forecasting project was specifically studying wind patterns in the tropics.

This is one of the most exacting applications of satellite data, demanding great precision in finding the position of the clouds (and therefore in finding the position and orientation of the satellite) and following the motion from one image to the next.

a British official associated with the programme said.[6]

The satellite was also to be used for gathering information on the temperature of the ocean surface by means of infra-red images, and use the same infra-red facilities to track invisible vapour clouds in the upper atmosphere, studying wind speeds and direction. The water vapour research was a unique feature on Meteosat which distinguished it from US weather satellites already in orbit.

The satellite functioned well for two years, transmitting images every half-hour and providing some stunning views of the Earth from 36 000 kilometres out in space, swirled in clouds and floating in a black void. A problem on the satellite put an end to the imaging mission at the end of the 1970s, when a small resistor in a power supply protection circuit malfunctioned, but the orbiter continued to collect data until the mid-1980s, paving the way for the more sophisticated satellites of the future.

Meteosat I was joined in orbit in June 1981 by Meteosat II. Its image-making equipment functioned well, replacing that lost on Meteosat I, but its data-gathering equipment failed. As a result the two satellites operated together, complementing each other's failure and providing a coherent picture of weather developments across a huge segment of the globe stretching from the Gulf to Brazil, and Scotland to the Cape of Good Hope.

The project achieved the considerable feat of staying within its original budget of 115 million accounting units (MAU) at mid-1971 prices, and ultimately led to the establishment of a new organisation, Eumetsat, to function as the administrator of an operational weather satellite system, in a similar role to that of Eutelsat for the telecommunications sector.

Apart from the main Meteosat programme, there were some related developments in Europe in the late 1970s in the field of meteorological satellites. The Italian national research institute (CNS) took the initiative in funding a satellite called Sirio-2, a geostationary orbiter which was designed to collect and transmit weather data through cheap ground stations. The project had clear applications for Third-World users. But it never became operational, due to an Ariane launch failure in September 1982 in which the satellite was destroyed when the rocket blew up.

The progress made by the space programme in Europe during the 1970s in pushing forward applications satellite programmes was immense, comparing very favourably with the previous decade when, plagued by technical and political problems, Europe had appeared to lurch from one crisis to another. At last scientists and engineers felt that the groundwork laid during the early pioneering years was beginning to show some tangible results, and they could look ahead to developing a generation of operational satellites which would serve Europe's needs and start to challenge the technological superiority of the more established space powers.

7 Ariane – The Commercial Challenge

THE KOUROU SPACE CENTRE

Sandwiched on a narrow coastal strip between 2000 miles of jungle on one side and 3000 miles of ocean on the other, the Kourou space centre is one of the loneliest and most extraordinary places on Earth. When French President Charles De Gaulle chose the place in 1964 as the launch site for French rockets, all that existed was a village of 600 people sited on rich farming land at the mouth of the Kourou river. Out to sea, an hour's boat trip from shore, lay Devil's Island, the notorious French penal colony where convicts had been incarcerated since Napoleonic times, tormented by the hallucinating heat, the humidity and the insects. As the space centre grew, first to 5000 technicians, then 10 000, then close to 15 000 by the late 1980s, life changed radically for the rest of the population. Up the coast from Cayenne to the south, past Kourou to Saint Laurent on the Maroni river in the north on the Surinamese border, French and locals alike ate croissants and baguettes for breakfast, read *Le Monde* freshly delivered by Air France and watched the French overseas television service bringing the latest variety shows and news bulletins from Paris.

Kourou was ideal for a rocket launch site because its position, just six degrees north of the equator, meant rockets could take advantage of the slingshot effect of the Earth's rotation to put satellites into geostationary orbit with substantially less fuel than was needed from more northerly launch sites such as Cape Canaveral or Baikonour. The first launch from Kourou took place on the morning of 9 April 1968, when a 1.5-tonne Véronique sounding rocket blasted skywards with a scientific payload. Subsequent years saw increased activity as first the Europa, and then the Ariane programmes, got under way. And during this time an uneasy relationship developed between the space base and the local community, who depended on the French technicians for their livelihood yet who also resented the imposition of French culture on the indigenous way of life.

Indians in the jungle, who only a few years earlier had been wrenched from the stone age by the arrival of light aircraft and settlers, were now confronted with Ariane rockets thundering into the sky, further dramatic evidence of the cultural invasion that was going on around them.

Apart from the rockets themselves, there was also the large number of immigrants attracted to the country by the wealth generated by the space programme. Up to 85 000 people lived along the coastal strip, of whom nearly 20 000 were Haitian labourers, several thousand more were

49

Brazilians and an estimated 8000, or 10 per cent, were Surinamese refugees fleeing the civil war in their native country just to the north.

Growing fear of attack from Surinam, or from other quarters, as the Ariane programme developed induced the French authorities to step up security at the Kourou space centre, first with an Atlantique naval patrol aircraft and subsequently with Crotale anti-aircraft missiles and 20 milli-metre anti-aircraft guns. The base was already playing host to the French foreign legion, whose members were to win unwelcome notoriety in the mid-1980s as the result of a rampage through Kourou's red light district ending in a running fight with locals and the death of a legionnaire.

Ultimately it was not so much protection that the space centre needed, however, but a better rapport with the extraordinarily varied cultural climate in which it found itself, and which it had helped to generate.

Among the more extreme examples of cultural transplantation, even more incongruous than that of Parisian technicians calmly munching their croissants, were a group of H'mong tribespeople who had been forced from their homes in the jungle around the Thai/Laotian border during the war in neighbouring Vietnam and Cambodia, shipped through refugee camps in Southeast Asia and France and had finally been offered accom-modation in two specially-constructed jungle settlements on the Guianese coast. They were careful to preserve their lifestyle as intact as possible, despite hostility from local people who resented their success at growing fruit and vegetables for market and questioned the value of transplanting them more than half-way across the globe. The H'mong had to compete with Brazilians, Colombians and Peruvians for a stake in the local economy, which was being increasingly dominated – perhaps inevitably – by the French.

The Kourou space centre, while an anachronism, was an un-doubted boon to the local economy. But while the local Guianese benefitted through construction work on the launch site and in providing services for the growing number of French moving to Kourou, it was also undeniable that qualified jobs on the space programme itself were virtually impossible for the local population to obtain. In the final analysis they were largely spectators, willing or unwilling, to the transformation of their once-sleepy backwater into one of the world's premier space ports.

ARIANE TEST FLIGHTS

After the mixture of euphoria and relief that had greeted the success of the first Ariane launch in December 1979, the second flight was approached in a spirit of cautious optimism. The rocket's three stages, including its complex cryogenic third stage, had all performed flawlessly, and there was little reason to doubt that they would do so again.

Five months after the inaugural flight, on 23 May 1980, a second Ariane rocket stood on the launchpad at Kourou, with two satellites in its cargo bay, a West German scientific satellite somewhat dramatically named Firewheel and an amateur radio satellite, Oscar-9. The launch was planned for 08.30 local time, but the countdown was plagued with difficulties. After one interruption due to technical problems less than a minute before the scheduled lift-off time, the count was resumed, only to be suspended again when a violent storm passed over the launch site, darkening the jungle landscape. Computer problems caused two more interruptions once the weather cleared and it was not until the fifth attempt that the rocket finally lifted clear of the pad and, climbing out across the ocean, disappeared into cloud.

The first hint of trouble came 72 seconds into the flight as instruments in the Jupiter control room 12 kilometres from the launch site indicated that one of the four Viking first-stage motors had shut down well ahead of schedule. As the rocket became less stable and its acceleration dropped, the performance of the other engines was impaired; 104 seconds into the flight pressure dropped sharply in two of the remaining three engines, and four seconds later the final one shut down as the rocket was torn apart; 112 seconds into the flight the mission controller reported loss of contact with the launcher.

> In an incredible silence everybody watched the display panel where, after 'lift-off', nothing else appeared. The ends of the two trajectory plotters fell back to zero ... A picture appears on the cinetheodolite screens: a twisted piece of steel falling towards the sea, turning over and over, like a dead leaf...

Peter Creola, chairman of ESA's Ariane Programme Board, wrote later.[1]

The inquest into flight LO2 started immediately, as helicopters and a fleet of boats headed for the crash site to fish as much wreckage as possible out of the ocean. At the same time Ariane officials started a meticulous examination of the computer tapes of the launch, hunting for clues to the accident. It was not long before the cause became apparent. High-frequency pressure oscillations in D engine started as early as 4.4 seconds after first stage engine firing, corrected themselves, recurred briefly at the 28-second mark and then worsened from 63.8 seconds into the flight, throwing the rocket into a slow roll. Despite this it continued flying close to its planned trajectory until two more engines failed at the 104-second mark, spinning the rocket around on itself and subjecting it to massive aerodynamic forces which ripped it apart.

The search for wreckage lasted more than three weeks, and it was not until Monday 16 June that divers succeeded in recovering the remains of D engine which had caused the crash. This confirmed what the investigators had already deduced. The injection system had melted under the stress of the pressure oscillations, and the combustion chamber had been torn open.

Despite the seriousness of the setback, the Ariane team was at pains to stress that launches would resume once the fault had been corrected. As an official ESA bulletin said three months later:

> The LO2 failure does not call the continuation of the programme into question. When the cause of the engine failure has been identified and the necessary corrective actions taken, the programme authorities will conduct the other two planned qualification firings.[2]

A year of intensive tests followed on a modified design for the engine to prevent a recurrence of the crash, 169 tests totalling nearly three hours of engine burn, an exhaustive rehearsal for an in-flight burn lasting just two minutes 25 seconds. Launch resumption was at first scheduled for March 1981, then pushed back three months to permit further checks. Finally on Friday 19 June, more than a year after the ill-fated LO2 flight, Ariane LO3 was ready to fly. The rocket had two satellites on board, the second European weather satellite Meteosat-2 weighing 700 kilos and the slightly smaller Indian experimental communications satellite, APPLE, built for the Indian Space Research Organisation (ISRO).

After two minor interruptions to the count-down, the rocket lifted off at 09.33 local time. The long months of meticulous preparation paid off, and the flight went perfectly. Once more the champagne flowed at a post-launch party for the workers at the Kourou space centre. This time they had not just the successful launch to celebrate but also news of the commitment of funds to a significant expansion of the base, with work due to start the following week on construction of a second launch pad close by the existing one, with a more sophisticated assembly arrangement capable of handling larger future-generation Ariane rockets.

The final qualification flight of the Ariane-1 rocket was planned for November of the same year, and after some slippage in the timetable eventually took place on the evening of 19 December. This was to be the first night launch of an Ariane, the timing determined by the very narrow 45-minute launch window of the satellite on board, the Marecs-A maritime communications satellite. The satellite, weighing just over one tonne, had been developed from the ECS series of telecommunications satellites and was owned by the London-based International Maritime Satellite Organisation (Inmarsat).

Once again the flight went perfectly, ending Ariane's four-flight qualification phase with a respectable track record of three launch successes and one failure and clearing the way for the start of a four-flight promotional series of launches prior to full commercial flights getting under way.

PROMOTIONAL FLIGHTS

As a result hopes were high during the preparatory campaign for the fifth flight, which was to launch the follow-on marine communications satellite Marecs-B and an Italian experimental telecommunications satellite, Sirio 2, designed to beam meteorological information to Africa and help in the international synchronisation of atomic clocks. It was to be another night launch. Lift-off was at 23.12 local time on 9 September 1982, and as the rocket climbed away into the tropical sky everything appeared set for another smooth flight. The first-stage burn functioned normally, as did the second, and four minutes 45 seconds into the flight the cryogenic H-8 engine powering the third stage roared into life. Several more minutes passed before the ground station at Natal, Brazil following the rocket's path began showing it dropping away from its planned course. As it disappeared from the Natal screen, the ground station at Ascension Island failed to pick it up. Less than two-thirds of the way through its planned 15-minute burn, the third-stage engine had extinguished itself. Wreckage fell into the ocean 500 kilometres north of Ascension Island.

Once again the inquest mechanism swung into action, and once again it did not take long to identify the cause of the accident. The system pushing the liquid oxygen and hydrogen through into the third-stage combustion chamber had ruptured, depriving the rocket of thrust. Once again a redesign was necessary, this time incorporated into the updated HM-7B motor which was being built to replace the H-8 on the more powerful Ariane-3 rockets planned to come into service within a couple of years.

Once again no-one talked of abandoning the programme, despite the fact that the new setback had pushed the failure rate in the first five launches up to 40 per cent. One French newspaper, previewing the sixth, Ariane launch the following June, declared that

> beyond any shadow of doubt, a failure would not lead to the death of Ariane. The stakes are too high, as are the sums of money put in by European nations.[3]

The total cost of the Ariane development programme to date, including the costs of the design changes following the LO2 and LO5 failures, was Ffr. 5.12 billion (in 1982 prices). Two-thirds of this money had been put up by France, and despite the cost overruns the overall budget was still broadly in line with the level approved by European nations back in July 1973, taking into account their prudent allowance of a 20 per cent overshoot.

Ariane LO6, the second of the promotional flights, blasted off from Kourou right on schedule at 08.59 local time on 16 June 1983. The first- and second-stage burns passed without incident, but nerves were strained during the third-stage burn when the Ascension Island ground station temporarily lost contact with the rocket. It was quickly established that the fault lay with the ground radar rather than with the rocket, and the rest of the flight was successfully completed. Once again two satellites were placed in orbit, the European ECS-1 communications satellite owned by the European Telecommunications Satellite Organisation Eutelsat, and the West German amateur radio satellite Amsat P3-B.

The success of the sixth flight was of enormous importance to the Ariane organisation, both technically and commercially. It confirmed that the rocket was fundamentally sound, and that once more technicians had been able to overcome the disappointment of a launch failure, thoroughly analyse the cause and painstakingly redesign the relevant parts of the rocket to eliminate the fault. Commercially the flight was a big morale-booster, since Ariane had more than 30 satellite launches firmly booked. Aside from two further launches in 1983 it was planning no less than six flights during 1984.

Inevitably there was some slippage in schedules but the rockets, once launched, performed flawlessly. The final two promotional launches took place on 18 October 1983 and 4 March 1984, each placing an Intelsat V telecommunications satellite into orbit.

ARIANESPACE

Flight nine, which orbited a US communications satellite Spacenet-F1 on 22 May 1984, marked the first commercial flight by the newly-established Arianespace organisation.

Arianespace was set up by ESA to finance and operate the Ariane rocket programme and commercialise launch space on it. It was also given responsibility for operating the Kourou space centre. Its shareholders were 36 of the principal European aerospace firms, mainly those involved in actually building the rocket, as well as 13 major European bank groups and the French space agency CNES. It was formally established on 26 March 1980. The size of each country's shareholding was a reflection of its financial interest in the project, with French shareholders accounting for 58.5 per cent of the stock and West German 19.6 per cent. Groups of companies from other ESA nations held blocks of 3.6 per cent or less in the company. Its president was Frederic d'Allest, a leading French rocket scientist who had been involved in the space programme from the early days of the Diamant launcher.

The next flight, the tenth in the series, was a landmark for the organisation since it was the first flight of the more powerful Ariane-3 rocket. The principal difference between this launcher and the smaller Ariane-1 which had been used up until then was that the Ariane-3 had a stretched third stage and two solid-fuel strap-on boosters. The effect of the improvements was to increase the maximum mass the rocket could place into geostationary transfer orbit (GTO) to 2.58 tonnes from 1.75 tonnes. Flight 10, designated V10 (vol 10) in recognition of the French linguistic as well as technical domination of the Ariane programme, was successfully launched on 4 August 1984 and carried two European telecommunications satellites on board, ECS-2 for the Eutelsat organisation and the French satellite Telecom-1A, destined for both civilian and military telecommunications.

The next three flights were all Ariane-3 rockets, and all passed off without a hitch. Each carried two telecommunications satellites. V11, launched on 11 November 1984, placed in orbit the US Spacenet-F2 and Inmarsat's Marecs-B2, the replacement satellite for the Marecs orbiter lost in the LO5 failure. V12, on 8 February 1985, launched Arabsat-F1, owned by a consortium of Arab countries, and Brasilsat-S1, bringing satellite communications to the vast Brazilian hinterland for the first time. V13, which lifted off just three months later on 7 May, placed the US GStar-1 and France's Télécom-1B in orbit. And V14, reverting to an Ariane 1, blasted off on 2 July to send the European scientific probe Giotto on its journey towards a historic rendezvous with Halley's comet.

Ariane's run of nine consecutive successful launches firmly established its reputation as one of the world's leading satellite launch operations. And despite having had two launch failures in its 14-launch history, giving it a higher percentage failure rate than some of its US competitors, the memory of the early crashes was fading fast. So it came as a blow out of the blue when, on flight V15 in September 1985 the fifth Ariane-3 plunged into the Atlantic with the loss of both satellites.

ARIANE V15 FAILURE

From the public relations point of view the accident could not have been worse timed, since French President François Mitterrand, together with his Foreign and Defence Ministers, had decided to attend the launch en route to a nuclear test at the French Pacific atoll of Mururoa. As a result the launch attracted more than the usual amount of media attention in France. Mitterrand's trip had been planned in the wake of the sinking of the Greenpeace environmentalist ship Rainbow Warrior in Auckland harbour by French secret service agents the previous month. The ship's crew had been planning to lead a flotilla of protest boats into the nuclear test zone around Mururoa in a bid to stop the test until two bombs, planted by French agents, scuttled the ship, killing a Portuguese photographer on board and provoking an international outcry.

Mitterrand's visit to Mururoa in September was planned partly as a strong show of support for the nuclear test programme, and partly to draw attention to French technological prowess in general. As a result, Elysée Palace officials felt that a stop-over in Kourou to attend an Ariane launch would serve to further emphasise France's role at the forefront of technology, and perhaps divert some of the criticism that was still rumbling in the press after the Greenpeace affair.

The trip got off to an inauspicious start when coverage of Mitterrand's departure by Concorde from Paris' Roissy airport on the main French television evening news bulletin showed, not the futuristic monument to Anglo-French technology roaring down the runway as planned but a frustrated Mitterrand being forced to change planes after a technical fault grounded the original aircraft.

After Mitterrand and his team of ministers and officials had arrived in French Guiana, they continued to suffer transport problems. So by the time the President arrived in the Jupiter control room at the Ariane launch centre in Kourou on the night of 12 September, his officials were fervently hoping that things would start to go right. They didn't.

The rocket blasted off at 23.26 GMT, shattering the peace of the tropical night as it climbed out over the Guyanese coast. All went well at first, with the first and second stage burns taking the rocket out over the South Atlantic to an altitude of over 100 kilometres. But just over four and a half minutes into the flight the third-stage engine failed to ignite properly. For a while the rocket continued close to its pre-set flight path, but at around the seven-minute mark it started to fall away from its planned trajectory. Nine minutes into the flight the mission safety officer sent a signal to the rocket to self-destruct.

A stony-faced Mitterrand, who had followed the launch attentively on the giant control-room screen, rose and left the room, leaving the dismayed Arianespace officials to start the painful analysis of what had gone wrong. The President later returned to the control room to share his disappointment with the launch team. In the somewhat melodramatic words of French television commentator Paul Amar, he moved among them 'like a general consoling his troops after defeat in battle'.

It was clear that something had gone wrong with the third-stage ignition system, and a board of enquiry was set up to report on the accident by the end of the month. Some of the more superstitious Ariane scientists attributed the accident to the fact that Arianespace President Frederic D'Allest had not been wearing his traditional casual launch attire, featuring principally an open-necked tropical shirt, but had put on a tie in honour of Mitterrand's presence. The accident investigators found a more plausible explanation, however. The hydrogen injection valve in the third-stage engine had leaked, pre-cooling the system and preventing ignition.

Since the accident was attributed to a faulty component, rather than a design flaw, the delay to subsequent launches was minimal and flights were able to resume the following February, barely three months behind schedule.

Spurred on by the V15 launch failure, however, Arianespace announced that it would set up a self-insurance subsidiary, offering satellite owners with payloads on Ariane the option of contributing money to a pool for paying out to those losing satellites. The premium would be based on the statistical assumption of one failure every 15 flights. The move came in response to the soaring cost of space insurance following a string of satellite launch problems in 1984 and 1985, many involving the US shuttle. Insurance company losses in the space sector had jumped sharply to more than $500 million by the end of 1985, despite a rise in premiums to 20 per cent of the combined cost of the launch and satellite in 1985 from between 12 and 15 per cent in 1984 and between 6 and 10 per cent prior to that. Ariane's self-insurance initiative, which met a favourable response from the industry, had been planned before the V15 crash, but was given added urgency by it.

QUICK LAUNCH RESUMPTION

Launches started again quickly in February 1986 with an Ariane 1 flight carrying a French Earth observation satellite, Spot, and a Swedish satellite. The launch took place in the shadow of the Challenger accident, which had dominated headlines world-wide since the end of January, and so had the added significance that potential customers who had been planning to launch payloads on the shuttle were starting to look to Ariane as probably the only alternative for the immediate future.

The launch passed off without incident, with both satellites successfully placed in orbit, and attention switched to flight V17 planned for the following month, which was to inaugurate the new, second launchpad at Kourou, ELA-2. The new launchpad was an essential development for Arianespace, which had been restricted to a single launchpad, ELA-1, since the start of the Ariane programme but which now wanted to begin stepping up the number of flights per year from the annual rate of four achieved in 1984 and 1985 to eight or nine by the late 1980s. The new pad was also designed to cater for the larger Ariane-4 rockets, which were too tall for the ELA-1 pad to handle.

The whole of the new launchpad area covered about 50 acres and was designed to allow one rocket to be prepared for launch on the pad while another was being put together in a nearby assembly building. A kilometre-long rail-track linked the two sites, allowing the rocket to be transferred to the pad on a mobile launch platform. The design of ELA-2 was much more sophisticated than ELA-1, which only allowed one launch preparation at a time. This meant that, even after the phase-out of ELA-1 with the end of Ariane 2 and Ariane 3 launches, the launch rate capacity of ELA-2 would still be double that of ELA-1.

Unusually, because it was the inaugural flight from ELA-2 and also because of the three-month launch suspension following the failure of the

V15 mission, assembly of the Ariane-3 rocket for use in the V17 flight had been started around six months before the eventual launch. Lift-off was scheduled for mid-March, then delayed for a few days due to minor technical problems. The flight was due to carry a US telecommuications satellite, GSTAR II, for GTE Spacenet and a Brazilian telecommunications satellite, Brasilsat S2. The Brazilian satellite's forerunner, Brasilsat S1, had already been successfully launched a year earlier on Ariane flight V12, but for GTE Spacenet, a well-established customer, it was the first ride on Ariane since the loss of its Spacenet III satellite on V15 the previous September.

The countdown ran smoothly throughout the hours leading up to the launch on the evening of 19 March, with the new launch facilities apparently presenting no problems. Fuelling of the third stage was completed on schedule and all appeared set for lift-off when, eight seconds before ignition of the first stage engines, ground computers shut down the count. While the launch technicians realised what had happened, observers and journalists were unaware that the launch was being aborted and saw the digital clock on the giant screen at the launch control room ticking on down to zero. When zero was reached and nothing happened, there was a buzz of consternation and considerable confusion. In the moments that followed the launch abort a pressure problem developed with a helium tank in the third stage which required rapid intervention by technicians, but within a couple of minutes of the shut-down the rocket was fully stabilised.

Checks showed that the reason for the shutdown lay in faulty contacts on the relays of the cryogenic arms of the new ELA-2 launch tower, which were used to fuel the third stage and which were designed to swing away from the rocket just a few seconds before lift-off.

Repair work, combined with the necessity of draining the rocket of fuel and re-starting the count from the beginning resulted in a nine-day delay to the launch. When the rocket finally took off on 28 March, however, everything went well despite a slightly too powerful burn by the first and second stages, officially demonstrating the launch-worthiness of ELA-2 and apparently confirming that the problems of the Ariane V15 launch had been resolved.

The next launch programmed for May 1986 was from the old ELA-1 pad but in one respect was also a first for Arianespace, since it was the first time an Ariane-2 rocket was to be launched. The Ariane-2 was identical to the more powerful Ariane-3, which had already flown six times, but lacked the two small solid-fuel boosters which provided extra thrust. Because the Ariane-2 was capable of putting up to 2.175 tonnes into geostationary transfer orbit, compared with Ariane-3's 2.7 tonne capacity, the smaller rocket was ideal for launching large satellites in the two-tonne class which were too big to travel with a second payload on an Ariane-3. Principal customers for the Ariane-2 were the International Telecommunications Satellite Organisation (Intelsat), with its Intelsat V family of orbiters, and the French and German telecommunications authorities with the TVSAT / TDF series of direct broadcast television satellites.

Intelsat V F14, the fourteenth flight unit of the Intelsat V series designed to carry 15 000 telephone lines and two colour television channels, was booked on Ariane V18, with launch from Kourou set for the evening of 30 May. The initial count ran smoothly until 11 minutes before scheduled lift-off, when the Intelsat organisation requested a hold on the launch because of technical concerns over the satellite. Intelsat officials said that a problem had been discovered on a similar satellite being built at Palo Alto, California, and technicians needed to conduct urgent tests to see if the same problem could repeat itself on the orbiter aboard Ariane. The delay only lasted 50 minutes, and after the count resumed the rocket lifted off at 21.53 local time (00.53 GMT Saturday 31 May). The first and second stage burns went well but four and a half minutes into the flight, when the moment arrived for the third stage to take over, a problem in the ignition system prevented the engine from firing. On the screen in the Jupiter control room at Kourou, the green dot plotting the rocket's course fell slowly away from the planned trajectory. Just eight months after the V15 failure, the scene unfolding on the giant screen looked horribly familiar.

'The third stage has not ignited – the mission is lost', Arianespace president Frederic D'Allest announced as controllers exploded the rocket over the ocean. His tone of voice was very matter-of-fact and professional, betraying little of the emotion and disappointment that he must have been feeling. D'Allest had worked in the French space programme since the early 1960s and had fought for the Ariane programme during the dark days of the early 1970s when many in Europe doubted the wisdom of pouring fresh resources into rocketry after the Europa rocket fiasco. Now, after nine consecutive successful flights between June 1983 and July 1985, Ariane had suffered two failures in only four launches. It was soon apparent that the delay this time was likely to be longer than the three-month interruption to the launch programme suffered after the V15 failure.

ARIANE V18 ENQUIRY AND LAUNCH SUSPENSION

A commission of enquiry was announced to look into the cause of the crash, but Arianespace officials realised that this time it would not simply be a question of identifying the problem but also of instituting a painstaking and potentially time-consuming test programme to ensure that it did not recur.

We have a problem concerning the ignition sequence of the third stage. This sequence is very delicate. . . . It is clear that there is a great amount of work to do before we give the go-ahead for the next launch

D'Allest said.[4]

The Ariane failure added to a sequence of extraordinary space mishaps that had befallen western launch operators since the start of the year. First had come the Challenger accident in January, with the loss of the seven crew and the grounding of the shuttle programme. Then in April a Martin Marietta Titan 34D rocket blew up on launch at Vandenberg Air Force Base, California and on 3 May a McDonnell Douglas Delta rocket carrying a GOES meteorological satellite had to be destroyed after its main engine shut down 71 seconds into flight, the first Delta failure after a run of 43 successful launches which had started in September 1977.

The string of failures left western companies and government organisations, including the US military, temporarily with no means of launching satellites. The problem for the US military was particularly severe, since it had the bare minimum of only one operational KH-11 spy satellite in orbit and the Big Bird photo reconnaissance satellite aboard the Titan that exploded had been intended to supplement it in its monitoring.

As far as Arianespace was concerned, the Ariane failure came at a particularly inopportune moment. For while company officials had been careful not to capitalise on the shuttle accident by canvassing too openly for satellite cargoes already booked on the shuttle, or intended for it, they found that many satellite operators were taking the initiative in approaching them and making enquiries about launch space on Ariane.

The failure deals a blow to the plans Arianespace . . . had for filling the commercial payload void left by the US shuttle standdown. The Europeans must correct the persistent problems they have encountered with their liquid oxygen/liquid hydrogen third stage, both to get their own launcher programme back on track and to help restore confidence in the validity of the Western world's space programmes

US magazine *Aviation Week and Space Technology* wrote in an editorial after the accident.[5]

The enquiry into the V18 failure concluded that the ignition system of the third-stage HM-7 engine was at fault and recommended that it should be redesigned, both to make it more powerful and to spread the ignition jet better within the combustion chamber. Tests on the engine were conducted throughout 1986 and early 1987 at SEP's vacuum test bed at Vernon.

The interruption to the launches was very frustrating, both for satellite operators and Arianespace, especially the technicians based out in French Guiana who found that life on the edge of the jungle could quickly become rather tedious without the challenge imposed by the successive launch campaigns. Arianespace used the lull to initiate some maintenance and modification work on the launch site, while its sales team were kept busy fielding enquiries from satellite operators hoping for launch slots two or three years in the future. In November 1986 NASA chief James Fletcher said the US could be interested in using Ariane to launch some of its scientific satellites following the Challenger accident, while the US air force held tentative talks with Arianespace on the possibility of launching military global positioning system (GPS) satellites, used for weapon targeting and navigation, on the European rocket.

During 1986 Arianespace signed up a record 16 satellite launch contracts, half of them after the V18 failure which temporarily grounded the programme. 'It is a quite exceptional situation. We have benefited, so to speak, from the Challenger disaster', Michel Lavany, Arianespace's sales manager for Latin America, Asia and Italy, commented in January 1987.[6]

The list of satellites signed up, worth a total of Ffr. 5.4 billion, demonstrated the global nature of Ariane's market. They included two satellites from the European Space Agency, one the ERS-1 Earth observation satellite and the other the ECS-5 telecommunications satellite. France had signed up to launch the TDF-2 direct broadcast television satellite and Britain its Skynet 4B military communications satellite. There were two US television and telecommunications satellites, one joint US/Japanese communications vehicle, two purely Japanese communication satellites, three new-generation Eutelsat communications orbiters, one Intelsat VI orbiter, an Inmarsat II maritime communications satellite and an Indian television and telecommunications satellite, Insat IC. The surge of orders took Arianespace's total order book by the start of 1987 to 57 satellite launches, with a backlog of 40 satellites to be placed in orbit worth around Ffr. 12.7 billion. This meant that Ariane was fully booked until well into 1990.

SPACE DEBRIS AND SPIES

While the sales effort continued at full pace, and work progressed on the redesign and testing of the new third-stage ignition system, the attention of Arianespace officials was diverted by two entirely separate, and equally unexpected, events. The first was the explosion in space on 13 November 1986 of the third-stage booster from the Ariane-1 rocket launched the previous February with the French Spot satellite on board. The remains of the rocket had been circling the Earth in a polar orbit ever since. The explosion split the casing into several hundred pieces of debris scattered across hundreds of mile of space, where they continued to orbit, posing a potential hazard to other spacecraft. The incident served only to highlight the mounting anxiety of the US authorities over the amount of space junk in low Earth orbit, since even tiny fragments travelling at high speed could cripple a satellite or manned vehicle in the event of a collision. As many as seven stages from US Delta rockets were known to have exploded in space over the years, and the US believed other Ariane stages may also have exploded undetected in equatorial orbits, where the US was less capable of monitoring debris. American officials concluded the explosion was most likely to have been caused by particles of the third stage oxygen and hydrogen fuel, although this could not be proved.

The other diversion was if anything more bizarre and unexpected. On 19 March 1987 the French Interior Ministry announced that police had arrested six people and were seeking a seventh on suspicion of involvement in spying on the sophisticated cryogenic motor technology developed for the Ariane third stage at Société Européenne de Propulsion's Vernon test site. The six included two graduates of France's élite Ecole Polytechnique engineering college, Pierre Verdier and Michel Fleury, both aged 36, and a Romanian-born woman, Antonetta Manole, aged 41, all of whom worked for the National Statistics Institute (INSEE) in Rouen. Verdier's 31-year-old Soviet-born wife Lyudmila Varygina, Frenchman Jean-Michel Haury, temporarily based at the Vernon rocket site, Mme Manole's French husband and Philippe Maillard, a journalist, were also implicated. Arianespace president Frédéric D'Allest said on French radio that it appeared the spy ring had been passing information to the East bloc, and early in April France announced the expulsion of six Soviet diplomats from Paris, prompting tit-for-tat expulsions of French from Moscow.

The French press gave the story front page treatment and within a few days the case had developed all the ingredients of a spy novel. Pierre Verdier's lawyer Paul Walter told a radio interviewer that Mme Manole had denounced the group to the French counter-intelligence service DST because she had been jilted by Verdier, her former lover, when he decided to marry Lyudmila Varygina. 'Love triangle' headlines peppered French newsstands and attention began to focus on Lyudmila, who gave interviews to various newspapers following her release on bail in early April affirming that the charges were baseless and claiming that she had been ill-treated by the French police and prevented from contacting the Soviet embassy. The Soviet news agency Tass started a campaign to have charges against her dropped, while Soviet government spokesman Gennady Gerasimov held a press conference in Moscow with her sister Irina Sarkssian calling for the case to be resolved quickly. Some of the spy suspects were released from jail very rapidly and doubts about the solidity of the evidence against Lyudmila and the rest of the group surfaced in the French press shortly before Prime Minister Jacques Chirac visited Moscow in mid-May, with one French paper quoting Chirac aides as saying there 'isn't much evidence in the file' against Lyudmila.[7]

French newspapers reported that the government was anxious to prevent the case from souring Franco-Soviet diplomacy at a particularly delicate time in East-West relations and disarmament negotiations.

Confusion increased when the French news agency Agence France Presse (AFP) issued a story on 13 May quoting a confidential DST report as saying the expulsion of the six Soviet diplomats had not been due to any spying activities relating to Ariane, a report which prompted a rapid denial from the Justice Ministry questioning the interpretation the agency had put on events. However, the influential daily *Le Monde*, in its issue dated 15 May, joined the fray, asserting that

> above and beyond the legal and diplomatic complications there is the reality of the case, which appears to be more about a settling of scores between secret services – the DST scoring a victory over its adversary GRU, the Soviet military intelligence service – than about a spy affair with any real dimension to it.[8]

Le Monde suggested that the DST had used the Ariane spy allegations to force the expulsion from the Paris embassy of the assistant Soviet air attaché, Valery Konorev, who they were convinced worked for the GRU and whose recent transfer to Paris they had unsuccessfully sought to prevent. And the paper added that

> at the Interior Ministry the embarrassment is visible. The importance given to the affair at the time of the arrest of the group, thanks to selective leaks, bears little relation to the true significance of the case.[9]

On 21 October, seven months after his arrest, Pierre Verdier, the man at the centre of the allegations, was freed from jail on the orders of an appeal court, effectively closing a case which had generated considerable heat in Franco-Soviet relations while shedding very little light on Soviet spying activities in France and Soviet interest, if any, in Ariane motor technology. The main beneficiaries of the affair had been the French press.

The climate of espionage in the spring of 1987 was fuelled by two other events which, although minor in themselves, helped keep the Ariane issue bubbling in the newspapers. In early May an employee of a small local newspaper, *Le Courrier de Mantes* in the Yvelines area southwest of Paris, bought a used Citroen car from an employee at SEP and discovered, among a heap of old papers in the glove compartment, a 29-page document giving technical details of the oxygen injection and purging system on the Ariane rocket. The incident raised further questions about security at Vernon and was picked up in the national press.

Then, just over a month later, Ariane was back in the news when 59-year-old French engineer Pierre-Antoine Bourdiol was sentenced in Paris to five years jail for passing industrial secrets to the Soviet Union. His work at state aerospace group SNIAS, the forerunner of Aérospatiale, had involved quality control of semi-conductors used in the Ariane programme, as well as on missiles and telecommunications satellites, and he had admitted having contacts with four Soviet diplomats between 1970 and 1983. However, despite having access to secrets, he claimed in his defence to have only passed on information which was publicly available. He walked free from the court after the hearing, having already spent two and a half years in prison following his arrest.

It was ironic that, at the time the French press was splashing articles about alleged Soviet spying on Ariane, the European rocket, like the shuttle, was grounded while the Soviet space programme was notching up some of its most conspicuous successes. Soviet rocketry still relied on proven, tested but fundamentally not very innovative technology, meaning that there could well have been sophisticated space technology being developed in France of potential interest to Moscow. However, the *de facto* Soviet lead over the West in space exploits could not be ignored. This was dramatically driven home when on 3 May that year, while France was trying to defuse the spy affair ahead of Chirac's Moscow visit, the Soviet Union launched for the first time its giant new Energia rocket, the vehicle that would fire the new Soviet shuttle into orbit just 18 months later. And at the same time as Energia was making headlines, more than 300 kilometres above the Earth aboard the Mir space station cosmonaut Yuri Romanenko was half way through his record-breaking 326 days in orbit.

MOVES TOWARDS A LAUNCH RESUMPTION

Barely three months after the V18 failure, while tests on the third-stage injection system were still in progress, Arianespace issued a tentative launch schedule calling for a resumption of launches in February 1987. According to this timetable seven flights were planned for 1987, eight for 1988 and nine for 1989, more than Arianespace had ever launched before in a single year. However, the start-up date proved too ambitious, as it became clear towards the end of 1986 that additional tests would be needed on the critical injection system. 'We will not meet the February target, but the delay should only be a few weeks', Arianespace President Frederic d'Allest told journalists in December that year.[10]

Ariane officials said they expected the tests to take only an additional four to six weeks, but some of the press coverage following the announcement took a gloomy view of the prospects for an early launch.

> A new delay in Europe's Ariane rocket programme, suspended after a disastrous misfiring in May, will put new strains on Western rocketry to handle the backlog of commercial, scientific and military satellites waiting to be launched

a *New York Times* article said.[11]

As tests continued into the start of 1987 Ariane officials estimated that the V18 crash had cost Arianespace Ffr. 500 million in flight delays and related expenditure, with ESA member states bearing an additional Ffr. 700 million cost for research on the new ignition system. Test delays mounted and it was not until mid-September, more than 15 months after the V18 crash, that launches resumed.

ARIANE RETURNS TO SPACE

As always the French media coverage of Ariane was much greater than that in other European countries, since France was by far the largest funder of Ariane, paying for more than half the budget, and tended to view the programme as a symbol of national prowess as much as a venture in European co-operation. After more than a year of gloom-laden headlines about the Ariane programme, papers across the political spectrum were in no doubt that this was a launch which had to succeed.

> It is not just contracts which Ariane risks losing if it fails in its mission – the chaotic state of the American space industry gives some idea of the possible consequences of failure. But Ariane is also a rare example of successful European collaboration, and it is a vision of Europe which is also at stake in Kourou

the left-wing French paper *Libération* wrote in an editorial.[12]

The right-wing *Figaro* described the launch as 'crucial for the European space programme' and commented that 'the stakes are considerable . . . the very credibility of the European launcher is on trial'.[13]

When launch day, 15 September, arrived everything went well, however, and despite minor delays the Ariane-3 rocket lifted off at 21.45 local time (00.45 GMT) with two telecommunications satellites on board, Australia's Aussat K3 and Eutelsat's ECS4. After a flawless flight the two satellites were ejected into orbit. The main emotion in the control room after the flight was a sense of profound relief. 'We're all wearing smiles . . . a failure would have been a heavy blow', D'Allest told journalists.[14]

And in a later assessment of the long launch interruption d'Allest wrote that

> there is no doubt that the Ariane programme has come out of this phase even stronger than before, thanks to the superb work carried out since summer 1986 . . . It was absolutely essential for us to take the necessary time and make certain that flights could be safely resumed.[15]

Once back in action, Arianespace was anxious to press ahead with further flights to clear the tremendous backlog of satellites. The next launch, V20, took place just over two months later, when an Ariane-2 rocket placed the West German television satellite TV-Sat 1 into orbit. The flight went well but a cloud was cast over the mission because the satellite later had to be abandoned due to the failure of one of its solar panels to deploy, depriving it of power. Two further launches took place in early 1988, one in March and the other in May, placing three communications satellites in orbit – Spacenet IIIR for GTE Spacenet of the US, Télécom IC for France Télécom and Intelsat V F13 for the Intelsat organisation. By the spring of that year, however, attention was turning to the next milestone, the first launch of the more powerful Ariane-4 rocket.

ARIANE-4 ROCKET SERIES

Ariane-4 was in fact not one rocket but a family of six, a basic launch vehicle with various combinations of up to four solid and liquid fuel strap-on boosters enabling it to launch payloads ranging from 1.9 tonnes to 4.2 tonnes, compared with Ariane-3's launch capacity of 2.7 tonnes. It was designed to give Arianespace great flexibility in tailoring the rocket to the satellite cargo, enabling it to fine-tune its prices to keep it competitive with the growing competition from US, Soviet and Chinese rivals in the launch market. The new rocket had a much larger satellite cargo bay than the earlier Ariane-3 version and an extended first stage burning 220 tonnes of fuel instead of 143 tonnes. But it was basically the same in concept. The philosophy of the new rocket was to

> keep it simple and retain as many aspects of earlier versions as possible ... The satisfaction lies in having transformed the launcher without really changing it, that is in taking care to incorporate past experience

Roger Vignelle, director of the launcher programme at the French space agency CNES, commented.[16]

The first Ariane-4 flight, after some minor delays in early June, was set for 15 June, with three satellites aboard. The rides were free since the flight was effectively a test flight being conducted by the European Space Agency, not a commercial launch under Arianespace auspices. The three vehicles on board were ESA's Meteosat P2 weather satellite, a US communications satellite, Pan Am Sat 1, and an amateur radio satellite, Amsat IIIC. The 480–tonne, 60-metre-high rocket blasted off at 11.19 GMT and after a smooth flight injected the three satellites into orbit.

'This launch took place under the best possible conditions ... the trajectory was perfect', D'Allest announced in the Jupiter control room. At a Paris news conference immediately after the launch officials hailed the success of the test flight as a sign of maturity of the Ariane programme. 'Another dream of the Ariane programme has become a reality', Jorg Feustel-Buechl, director of ESA's space transport systems division, said, while French research minister Hubert Curien commented that 'today's launch was really very important ... at a time when the competition is becoming more active, it was essential for us to show that we have perfect mastery of launch techniques'.[17]

Ariane-4 was designed to replace the Ariane-3 and Ariane-2 rockets by the start of 1990, its six versions providing for single, double or triple satellite launches. The core vehicle without any boosters, called Ariane 40, was capable of lifting 1.9 tonnes into geostationary transfer orbit and was ideal for single satellite launches. The next version up, with two solid-fuel boosters, was called Ariane 42P, designed to launch 2.6 tonnes while the 44P, with four solid-fuel boosters, could lift 3.0 tonnes and the 42L, with two more-powerful liquid boosters, 3.2 tonnes. The most complex version with two liquid and two solid boosters, called 44LP, could orbit 3.7 tonnes while the most powerful, the 44L with four liquid boosters, could launch 4.2 tonnes. The solid boosters were built by Snia BPD of Italy, part of the Fiat group, while the liquid boosters were built by MBB/Erno of West Germany.

Following the test launch of the first Ariane-4, three more Ariane launches followed between July and October 1988 using one Ariane-3 and two Ariane-2s. The first commercial launch of an Ariane-4, on 11 December, placed in orbit the Astra direct-broadcast television satellite and a British military satellite while the following flight in January 1989, an Ariane-2 carrying an Intelsat communications satellite, marked the tenth successful launch since the V18 accident, breaking the previous longest-run of nine successful launches by Ariane which had ended with the V15 failure.

Five more launches followed in the space of barely six months, including the first flight of the most powerful version of the Ariane 4, placing a further eight satellites in orbit. The rockets all performed well although one satellite, ESA's Hipparcos astronomical satellite, never achieved its planned orbit due to the failure of its booster motor. The improved launch rate enabled Arianespace President Frédéric D'Allest to announce in June that his company had achieved its goal of attaining a rocket production and launch rate of between eight and nine flights a year – necessary to meet the demands of an order book which had grown to 34 satellites awaiting launch worth a total $2.3 billion to Arianespace.

Morale among Arianespace workers soared, with the success of the programme outweighing concerns over growing competition. Nevertheless, executives were conceding that the company was having to change its marketing strategy to face a quite different situation from that envisaged in the early 1980s.

When the Ariane programme was being developed, the assumption was that the principal, and possibly only, rival would be the US space shuttle. But the situation changed radically with the Challenger accident and

President Reagan's subsequent decision in August 1986 to bar NASA from bidding for new commercial satellite launch contracts for the shuttle, encouraging private US industry instead to build rockets to do the job. The shuttle was to be reserved for military payloads and scientific satellites. By early 1987 Arianespace was aware that the major US aerospace groups posed a bigger threat than other rivals in the commercial space business such as the Soviet Union, China and Japan. 'This is really where our competition is going to be coming from over the next few years', according to Arianespace executive Michel Lavany.[18]

The US manufacturers had been in the launch business much longer than Arianespace, building rockets under contract to NASA and the US Air Force, although during the 1980s they had seen their role superseded by the shuttle as the designated prime US launch vehicle. Once the White House gave the green light to constructors to go ahead and bid for commercial satellite launches, three main groups immediately entered the fray. The largest of the US rockets available for commercial launch was the Titan built by Martin Marietta, which could launch up to 5.4 tonnes into geostationary transfer orbit (GTO) and which had made over 130 flights by early 1988. A new version, the Titan 4 designed to launch even larger payloads in the early 1990s, made its first flight in June 1989. The smaller Atlas Centaur, built by General Dynamics, could put up to 2.4 tonnes into geostationary transfer orbit while McDonnell Douglas' Delta series could launch payloads of just under 1.5 tonnes. All had long track records and reliability rates of well over 90 per cent.

China, which only started offering commercial launches to western satellite operators in the late 1980s, was offering its Long March 2 and Long March 3 rockets for payloads of up to 2.5 tonnes while Glavkosmos, the commercial launch arm of the Soviet rocket programme, was beginning to offer flights on the Proton, a rocket with a success rate of over 90 per cent. Japan's smaller H-I could launch only 550 kilos, although its more powerful H-II planned for the early 1990s would be able to lift around two tonnes into GTO.

Arianespace launch prices varied little on a kilo for kilo basis with American competitors, although European officials said US manufacturers were tending to offer initially low launch prices to gain a market foothold. The main problem for Arianespace was the competition from the Soviet Union and China, which were able to offer launches at well below commercial prices because they were not seeking to make a profit.

There already is a clear policy in Europe and the US that encourages private companies to compete with their expendable launcher on a pure commercial basis, but this is totally incompatible with the policies of countries like the Soviet Union and China that offer their launcher at unrealistic 'political' prices

D'Allest was quoted as saying following the first Ariane-4 launch.[19]

Alongside the pricing issue was the reluctance of the US government to permit American satellites to be launched on Soviet or Chinese rockets for security reasons, fearing a leakage of high-technology secrets while satellites were in transit on Soviet or Chinese soil. By the start of 1989 the US and China had reached an accord permitting China to launch up to nine international satellites up to the end of 1994, an average of 1.5 a year over six years, providing prices were based on world market levels. Launches of US satellites aboard the Soviet Proton were still ruled out, however, and there was no sign of a framework agreement between Europe and the US on the ground rules for launch pricing sought by Arianespace.

Despite the tensions between rival operators in the increasingly competitive satellite launch market, however, the good news both for the rocket industry and satellite owners was that western launch capability appeared to have been restored to something approaching normality by the end of the 1980s after its disastrous sequence of accidents in 1986/87. After a period where rocket failures were rarely out of the headlines, satellite launches, to the relief of those involved in them, were once more becoming routine events.

8 Spacelab

The Spacelab project had its origins in the late 1960s, when Europe was still struggling to build a launcher that would fly and had barely started to orbit even the most basic satellites. It was at that time, with the US Apollo programme in full swing and the first moon landing imminent, that NASA approached the European space authorities to discuss its proposals for the next generation of space transport system. NASA scientists were already working on the concept of a re-usable space shuttle, to be launched on a rocket and land like a glider on a runway, and were interested in involving Europe in the project, partly in order to save costs.

ESRO, while enthusiastic to become involved, was anxious not to become a simple subcontractor on the shuttle, with its work anonymously integrated into an orbiting vehicle for which the United States would inevitably take credit. As a result of three years of discussions, starting in late 1969, European countries agreed to take responsibility for a single, identifiable project within the shuttle programme, the design and construction of a space laboratory to ride into orbit in the shuttle's cargo hold for a week or ten days at a time before returning to Earth for the experiments to be analysed.

GREEN LIGHT FOR SPACELAB

Formal go-ahead for the project came at ministerial level at the Brussels meeting of the European Space Conference in December 1972, and in September 1973 the nine countries which had decided to take part signed a memorandum of understanding with NASA concluding the deal. From the start West Germany was the most enthusiastic, seeing the potential space offered for expanding scientific research into completely new areas, and agreed to contribute 54.9 per cent of the cost of Spacelab by itself, estimated initially at around $750 million but later to climb to around $1 billion. In return West German firm VFW ERNO (now MBB-ERNO) became prime contractor on the project, responsible for its overall management as well as system engineering and the final assembly of both the trial engineering model and the laboratory itself. AEG-Telefunken won the contract to supply the electric power distribution while Dornier was given responsibility for life-support systems and control of the laboratory environment. Italy was the second largest contributor, putting in 15.6 per cent of the cost. This reflected the fact that Aeritalia was to build the module containing the laboratory. France was the third-largest contributor, with Matra taking over the command and data management of the laboratory, while Britain was the fourth-largest contributor with 6.5 per cent, British Aerospace taking responsibility for design and construction of the pallet, an annex to the main module on which experiments could be placed to be exposed to the harsh space environment.

Other countries involved in the project were Belgium, Spain, the Netherlands, Denmark, Switzerland and Austria, each contributing less than 5 per cent to the cost. In addition to the MBB-ERNO group, whose two constituent parts had actually competed with each other for the Spacelab contract prior to their merger, there were ten contractors on the programme and a further 36 subcontractors.

Industrial contracts were awarded during 1974 for a programme which was scheduled to reach maturity by the early 1980s. An initial review of Spacelab's design was conducted during the first half of 1976 and a final design review took place in early 1978. During the development phase engineers and scientists had to keep adjusting their plans to take account of changes in specifications of the US space shuttle, since it was necessary to ensure that the two systems were totally compatible with each other.

PROJECT DESIGN

The Spacelab design actually comprised two quite separate concepts, one a pressurised module in which astronauts could work, supervising experiments which required human intervention or monitoring, and one the so-called pallet, which was actually a U-shaped configuration of panels exposed to space on which a variety of experiments could be placed, or from which observation of the space environment or astronomy could be carried out. Both segments were designed to fit in the cargo bay of the shuttle, with the cylindrical module connected to the cockpit section of the shuttle through a pressurised access tunnel. The module, four metres in diameter, came in two sections, each 2.7 metres long, permitting either a smaller core laboratory or a larger version with considerably more working space to be taken into orbit, depending on the demand for experiments. Because all the essential systems for supporting both the astronauts and the experiments were inevitably located in the core module, the single-module configuration left only eight cubic metres available for experiments, while the double-module configuration made a total 22.6 cubic metres available, nearly triple the working space.

The total length of the double-module configuration, including cones anchoring it to the hold, was seven metres, while the single-module was 4.3 metres. In comparison the cargo bay was 18.3 metres long and 4.6 metres wide. The area not taken up by the pressurised module sections of Spacelab was devoted to the pallet, at the far end from the cockpit, and the access tunnel, at the near end. The pallet, shaped to fit snugly into the walls and floor of the shuttle cargo bay, was three metres long with a total width across the cargo bay of four metres. It was possible in theory to fly up to five pallets side-by-side in the cargo bay without the module section at all, with key instruments housed in a cylindrical container called an igloo on one of the pallets.

Some of the experiments could be monitored from the cabin of the shuttle, with astronauts checking equipment on closed-circuit television cameras and using computer terminals to call up data displays. Inside the laboratory module were a series of racks, built either singly or in double format, 56 and 105 centimetres wide respectively, and 76 centimetres deep, designed as much as possible to take standard laboratory equipment in order to make experiments easier to plan and to hold down costs. There was a work bench for the crew, hand and foot holds and a range of equipment to prevent materials from floating freely around the module.

The standard temperature within the module was set at a pleasantly warm 22 degrees Celsius, with 50 per cent relative humidity. Electrical power available to the Spacelab module averaged between three and four kilowatts an hour, peaking at between eight and nine, with slightly more power available to the external pallets. Each pallet had 33.5 cubic metres available for experiments, compared with the 22 cubic metres available in the long version of the module. The pallets could each hold three tonnes-worth of equipment for experiments, and were designed to be controlled through a computer keyboard either from the Spacelab module or the cabin of the shuttle.

The internal design of the Spacelab module was very simple, but also, from an aesthetic point of view, rather boring, as astronauts who flew in the laboratory later commented. Working in the module was rather like spending a week inside a rail freight container, its walls lined with standardised 48-centimetre experiment racks, clinically arranged to make maximum use of the limited space available. However, its great advantage was that, being fully pressurised, with a carefully-regulated living environment, astronauts could dispense with the clumsy suits that had character-ised the early space missions and work in shorts and shirt-sleeves, making for an atmosphere of informality.

SALYUT SPACE STATIONS

While European scientists and engineers worked on development of Spacelab through the mid and late 1970s, both the Soviet Union and the United States were separately building up experience in constructing and operating the first space stations. The Soviet Union, having lost the race to the moon in the 1960s, had gained a head start in the battle to be first in orbit with a space station. On 19 April 1971 the Salyut 1 station was launched, a historic step that was soon to be overshadowed by unforeseen difficulties and tragedy. The first crew to attempt to visit the station later the same month managed to dock with difficulty but were then forced to cut short their flight and returned to Earth without ever entering the station. The second crew on board arrived on the Soyuz 11 spacecraft for a 23-day stay during which they conducted Earth observation and life science experiments and carried out astronomical observations. The working and sleeping quarters of the space station were less cramped than earlier spacecraft, with a diameter of nearly four metres, while the whole complex, including the Soyuz craft, was around 20 metres long. The mission went remarkably smoothly, in view of its pioneering nature, until the time came to return to Earth, when an accident on re-entry led to the sudden depressurisation of their capsule, instantly killing all three crew.

Salyut 1 was never used again, and its successor Salyut 2 disintegrated after launch. Salyut 3, which was orbited in June 1974 and visited for 15 days the following month by a two-man crew aboard Soyuz 14, was the first example of a space station being used for military observation purposes, while its successor Salyut 4, launched in December of that year, was used for scientific experiments and observation. Salyut 5 went into space in June 1976, followed by Salyut 6 in September 1977. Despite several mishaps, technical problems and near disasters in flight during the course of the Salyut programme, the Soviet Union steadily built up its expertise during the 1970s in manned space laboratory techniques. By the end of the decade cosmonauts were clocking up several months at a time in Salyut, preparing the way for the series of long-duration flights in the 1980s.

SKYLAB

The US attempt to place Skylab, its first laboratory, in orbit nearly failed dramatically at the very start. Barely a minute after launch on 14 May 1973 aboard the same type of Saturn V rocket that had taken men to the moon, a thermal shield and one of the laboratory's solar panels were ripped away. When it reached orbit its second solar panel, which had become jammed in the launch accident, refused to unfold, leaving Skylab without an energy source and unprotected from the direct heat of the sun. The project, on the verge of disaster, was saved when a crew of three led by Charles 'Pete' Conrad managed to dock with the crippled spacecraft later in the month, erect a heat shield and extend the jammed solar panel during a spacewalk. In the debate that was to erupt in later years on both sides of the Atlantic over the usefulness of manned space flight, and the extent to which men could be replaced by robots in space laboratories, the rescue of Skylab after its disastrous launch was held up as a prime example of the role men could play.

It was really the men who saved Skylab. No robot could have constructed
the solar shield, it was an astronaut who deployed the solar panel

ESA official Jean-Jacques Dordain said.[1]

That mission, and two subsequent flights later in the same year to
Skylab, carried nearly 300 experiments, ranging from studies of live spiders
and fish in space to Earth observation and astronomy. Around 40 000
photographs of the Earth were taken from Skylab's vantage point 435
kilometres up, and more than 180 000 pictures of the sun. But once the
third crew had left the station in February 1974 it was abandoned, its orbit
slipping lower and lower due to its restricted power supply until it finally
burned up in the atmosphere in July 1979.

SPACELAB PREPARATIONS

During the second half of the 1970s, while the Soviet Union was developing its Salyut technology and the US manned space programme passed
through a relatively sterile period ahead of the first shuttle launch,
European scientists worked on Spacelab. In early December 1980, barely a
year after the first Ariane launch and at a time when European space
projects were forging ahead on several fronts, the engineering models of
two Spacelab modules and three pallets arrived at Cape Canaveral from
their final assembly hangars at the MBB-ERNO facility in Bremen, West
Germany. The engineering models were intended to be identical to the
final version, only not fully tested for their ability to withstand the rigours
of launch, flight and the space environment.

NASA was formally responsible for the testing, launch preparation and
in-flight operation of the Spacelab, but ESA maintained a transition team
in Florida to help the NASA scientists throughout the long pre-launch
testing period and liaise with manufacturers back in Europe. During the
testing of the engineering model a problem arose with the freon coolant
planned for cooling the laboratory's computer system, which NASA feared
could be dangerously toxic, and a replacement was found. The ESA
bulletin of August 1984, reviewing the preparations for the first Spacelab
flight, recalled that

> there were also numerous problems related to electrical ground support
> equipment (EGSE) hardware failures, servicer gas line leakages, initial
> test configuration errors, procedural errors and hardware unfamiliarity,
> all of which caused test interruptions.[2]

None of the problems proved insoluble, however, and a further big step
towards launch was taken with the arrival of the first flight unit of the
Spacelab in Florida by early 1982. Initial tests proceeded smoothly through

the spring and summer, until the time came to perform a rigorous examination of the module subsystem, complete with redesigned avionics units.

The results of the test were dramatic in that, at the end of September 1982, just one year before the scheduled launch, several major problems had to be faced ... a mixture of general design deficiencies and individual box failures which in total represented a breakdown of the entire command and data management system

the ESA bulletin said.[3]

The setback would have been more serious had the whole shuttle programme not itself been delayed by a series of technical problems, resulting in launch delays which had already pushed back the date of the first Spacelab flight. In fact an extraordinary burst of activity ensured that deficient equipment was sent back to Europe, corrected, tested and returned to Cape Canaveral by the start of 1983, barely three months after the problems had come to light.

With about six months to go before the launch of the first mission, Spacelab-1, tentatively scheduled for late September 1983, tests were carried out on the cargo of experiments that were to fly on board. About one-third of the planned total mission time of 215 hours was simulated in a ground test during the spring, followed by tests to check that the finished Spacelab model was totally compatible with the shuttle in which it was to fly. Although all the tests were proceeding on schedule, the September launch date had to be put back to late October after the first Tracking and Data Relay Satellite (TDRS), a vital part of the communications system for the space shuttle, was left drifting in a useless orbit following a launch accident and had to be rescued by intricate ground control manoeuvres.

On 16 August the Spacelab flight module was lowered into the cargo bay of the space shuttle Columbia for another series of tests, and taken the five kilometres out to the launch pad on the giant NASA tractor platform at the end of September. The pre-launch checks that followed on Spacelab and the orbiter went smoothly but a problem had arisen during the previous launch of the shuttle, mission STS-8, at the end of August, concerning the nozzle of one of the giant solid rocket boosters. The problem, which could have destroyed the spacecraft in mid-flight, had come to light after the re-usable boosters were recovered from the Atlantic after the launch. As a safety precaution it was therefore necessary to replace parts of one of the boosters on the Spacelab mission, forcing launch officials to roll the complex launch ensemble of shuttle, external tank and boosters back off the pad for dismantling.

Despite a series of unrelated problems subsequently arising with the shuttle, all of which added to the frustration of the Spacelab team who could do nothing to help and simply had to wait out the delays, 28 November was chosen for the new launch attempt. On 8 November

Columbia was rolled back to pad 39A, the boosters having been cleared for flight, and the final launch preparations began.

The two-month slippage in the launch date meant that seven out of the 72 experiments on board would produce significantly less data than originally foreseen, and these experiments were therefore allocated reflights later on subsequent Spacelab missions.

NASA and European officials decided to press ahead with the flight, however, concerned that further delay could be detrimental to an ever wider range of experiments.

INAUGURAL LAUNCH

Launch day, 28 November 1983, was sunny and clear and the crew were in characteristically high spirits at the pre-launch briefings. Commanding the flight was John Young, at 53 one of the most experienced astronauts in NASA. A US Navy pilot, his space career had taken off nearly 20 years earlier when he made three Earth orbits in the Gemini 3 capsule in March 1965. This was followed by a longer trip aboard Gemini 10 in July 1966, a swing around the moon on Apollo 10 in May 1969 and command of the Apollo 16 mission to the moon in April 1972, including three days on the lunar surface. Earthbound for nine years during the slowdown in the US manned space programme, he then splashed spectacularly back into the headlines by commanding the first shuttle flight in April 1981. With the Spacelab mission he became the first man to make six space flights.

The other crew-members on the flight were all making their first space flight, with the exception of 53-year-old mission specialist Owen Garriot, who had been into space ten years earlier on the second of the manned Skylab flights. His fellow mission specialist, Robert Parker, had been part of the ground team on the Skylab programme. Young's pilot for the mission was Brewster Shaw.

Mission specialists are scientists recruited by NASA to be part of the permanent astronaut core, trained in handling various scientific aspects of space flights and capable of working on several different missions. Payload specialists are scientists recruited for a specific flight.

For the first Spacelab flight, in addition to the two pilots and two mission specialists there were two payload specialists, one European and one American, reflecting the roughly equal split in the experiments on board between ESA and NASA. West German Ulf Merbold, a 42-year-old solid-state physicist, had been on the ESA staff for nearly six years training for the flight while his US colleague, MIT biomedical engineer Byron Lichtenberg, had been an air force pilot in Vietnam and was at 35 the youngest man on the crew.

Both payload specialists had ground back-ups who followed an identical training course and could have replaced them on the flight in the event of illness or other problems. NASA back-up was Michael Lambton while Merbold's back-up was Dutch nuclear physicist Wubbo Ockels, who by the time of launch had already been selected to fly aboard the German D1 spacelab mission a couple of years later.

Merbold's presence on the flight made history for the European Space Agency, signalling its entry into the era of manned space flight. A Frenchman, Jean-Loup Chrétien, had already been into space the previous year on the Soviet Soyuz T-6 mission to the Salyut-7 space station, but that flight took place under a bilateral accord between France and the Soviet Union and was quite independent of ESA. Merbold was the first non-American to fly on the shuttle, and as a German a particularly appropriate crew member since West Germany had been by far the most enthusiastic backer of the Spacelab project, putting up more than half the funds and taking the lead in its design and construction.

The crew boarded the shuttle Columbia, the original shuttle orbiting vehicle now about to make its sixth flight, at 08.10 EST on 28 November, just under three hours before launch. A comprehensive pre-launch run-through of all systems 20 minutes later uncovered no problem and the count-down progressed to a scheduled hold shortly after 10.30. At 10.51 it resumed with nine minutes to go to launch; 31 seconds before blast-off Columbia's computers took over from those on the ground and at 11.00 the orbiter's main engines fired; 6.6 seconds later the giant solid rocket boosters ignited and the shuttle climbed away from pad 39A, rolled over onto its back and headed up the Florida coast.

> After so many disappointments over postponed launch dates, the event itself came as a spectacular success. Unlike the initial slow climb of the larger rockets, the shuttle seemed to spring away from its pad

ESA director-general Erik Quistgaard reported.[4]

Columbia, going supersonic within 16 seconds of launch, was in orbit eight minutes later and Spacelab's systems were activated two and a half hours into the flight. There was a dreadful moment when the crew attempted to open the hatch leading into the laboratory from the shuttle deck and found it jammed, potentially turning the mission into an embarrassing disaster. But the problem was overcome, and within three and a half hours of launch Garriot had floated through the tunnel into the laboratory to start up the experiments.

WORKING IN ORBIT

The astronauts were divided into two shifts, each of 12 hours, to keep work going round the clock during the ten-day flight. Each shift comprised one pilot, one mission specialist and one payload specialist to ensure smooth functioning of all systems. Garriot worked with Lichtenberg and Shaw while Merbold took his turn of duty with Parker and Young.

After the years of preparation, there was a sense of elation among scientists back on the ground that the flight was finally under way. But despite the minutely-planned flight, some things did not go quite as expected. The laboratory was full of sophisticated equipment, but no-one had thought to pack a sponge. As a result spilt liquid from some of the experiments caused quite a problem, and used up precious supplies of tissues.

> We had seven years to prepare the first flight of Spacelab ... I think maybe we had too much time. We practised the experiments hundreds of times, but they still worked out differently

ESA's Jean-Jacques Dordain said.[5]

The experiments were divided into the five general areas of atmospheric and solar physics, astronomy, life sciences and Earth observation. This wide range of experiments was possible because of the presence on the first flight of both the pressurised module and the unpressurised pallet, exposed to space at the back of the cargo bay.

During the first few hours of the mission the crew concentrated on performing experiments on themselves, closely monitoring their own bodies' adaptation to weightlessness. On both the Spacelab flight and other US and Soviet manned space station missions, the study of the human body was a key feature. Not only did weightlessness provide a fresh angle from which the body's general physiology could be studied, but the study of man in space was essential for the preparation of longer-term space flight and perhaps eventual interplanetary missions. While the Soviet Union had gained by far the greatest experience in this field, due to the resources it had poured into its Salyut and Mir space station programmes, the Americans and Europeans were determined not to be left too far behind.

Another key role of the first Spacelab flight was to monitor and analyse the performance of Spacelab itself as an orbiting laboratory. The system, despite having been subjected to a massive array of tests at every stage of its development, had been designed and built on Earth, and it was important to be sure that its behaviour in orbit was in accordance with predictions. With the experiments on board of such extreme sensitivity, any variation in laboratory environment had to be meticulously recorded in order that data did not become distorted.

As a result the orbiter was put through a wide variety of manoeuvres during the flight to expose the module and pallet to the largest possible range of temperatures and conditions, with in total more than 200 adjustments of position either because the experiments required them or for the purpose of testing the laboratory's functions.

To initiate the sequence, Columbia was positioned with its tail facing towards the sun, shielding the payload bay from direct radiation and creating the coldest possible environment for the laboratory. Most experiments were unable to operate under the cold test conditions, but it formed an essential part of the process under which Spacelab could be certified fully operational.

Various astronomy, plasma physics and life science experiments were able to continue during the period of the cold test, but once it was drawing to a close on the third day of the flight other experiments in the fields of atmospheric physics, fluid physics and material sciences were started up.

Some experiments had gone more slowly than anticipated on the first day of the flight, as the crew adapted to weightlessness and the novelty of orbit after the years of simulations. There was also a problem with a piece of equipment called a remote acquisition unit (RAU), which was an important link to the experiments on the pallet, although a back-up was available.

OPERATIONAL PROBLEMS

Nevertheless, once the working rhythm had been established everything worked well until the fourth day of the mission, when a problem arose with the High-Data-Rate Recorder (HDRR), a piece of equipment which was used for storing data that could not be immediately transmitted back to Earth.

Spacelab had been designed to transmit experimental information back to Earth via two NASA tracking and data relay satellites (TDRS), but due to the problem with the launch of the first of the pair, the second had not been orbited by the time Spacelab flew. As a result up to half the data from the mission had to be stored on the HDRR for later transmission to Earth. The HDRR went out of action for around 11 hours after suffering a surge in electric current through its tape drive motor, but an emergency back-up system existed. Robert Parker, the US mission specialist, was able to resolve the problem, so preventing large quantities of hard-won data from being lost.

Other difficulties which the astronauts were also able to resolve included the jamming of the infra-red metric camera and a power problem in a rack housing materials science experiments.

The arguments against the need for man's presence on board have been very convincingly refuted ... perhaps the most astonishing vindication occurred with a fluid-physics experiment, for which the principal investigator on the ground and the Spacelab crew were able to redesign the approach during the mission, thereby enhancing the value of the data recovered

Quistgaard observed.[6]

SCIENTIFIC RESULTS

The experiments in general were enormously successful, with discoveries being made in a variety of fields. The atmospheric physics experiments turned up new information on the density of heavy hydrogen in the upper region of the atmosphere known as the thermosphere, more than 85 kilometres above the Earth's surface, and the existence of carbon dioxide and water in the layer just below called the mesosphere. Studies were also carried out on the ozone layer, whose vulnerability to pollution had begun to cause serious concern beyond the environmentalist and scientific circles which first drew attention to the problem.

Pollution in general was becoming a constant theme of Earth observation from space, both as a target for formal experiments and as a phenomenon easily observable by astronauts gazing out of their windows in off-duty moments.

If you take a picture of anywhere in the world and compare it with pictures taken by astronauts in 1974 (when Skylab was in orbit), there is a very distinct deterioration in clarity. The horizontal spread of dirt is very rapid

one astronaut said.[7]

Experiments in space plasma physics on board the Spacelab 1 mission studied the immediate environment of the shuttle in orbit and the spacecraft's effect on its own surroundings, while astronomical observations and studies of the sun were also carried out. Life science experiments included close monitoring of the crew themselves, including their reflex actions and the way the body adjusts its balance mechanism in a weightless environment. In the area of materials sciences a large silicon crystal was grown for the first time in orbit, with the aid of a mirror heating facility. Experiments on liquids included investigations into the nature of convection and studies of the way liquid behaves in zero gravity, aimed partly at improving the design of spacecraft cooling and fuel systems.

Apart from the serious scientific work, there was also some time for public relations and media hype, with NASA organising a television link-up between President Reagan, West German Chancellor Helmut Kohl (who was on a visit to Athens at the time) and the Spacelab crew at the end of their seventh day in orbit.

BACK TO EARTH

Columbia had been due to return to Earth on its tenth day in orbit, but there was sufficient fuel and back-up laboratory supplies to enable an extra day to be added to the flight to allow more time for experiments. When the time finally came to fire the orbiter's engines for re-entry, a problem with two of its flight computers forced a postponement of the manoeuvre. It took another five orbits before mission commander John Young was able to swing the shuttle into its descent curve, blazing through the upper atmosphere at more than 20 times the speed of sound before making its final approach low over the Californian coast to land at Edwards Air Force Base.

The first flight of the Spacelab had been the longest shuttle mission to date, totalling ten days, seven hours and 47 minutes. Spacelab had been continuously in operation for just over nine days and 13 hours of that time, and both NASA and ESA were jubilant over the scientific results achieved from the experiments and the smooth functioning of the laboratory itself.

> The number of technical problems has been remarkably small for a first flight. The quiet efficiency with which they were mastered speaks volumes for the preparations and documentation without which a mission of this complexity cannot be mounted or fully exploited ... we are only now realising what possibilities and potential have been opened to us by Spacelab

Quistgaard wrote.[8]

Spacelab, like the shuttle itself, was designed as a re-usable system to make frequent space flights and open up the possibility of research in orbit to as wide a scientific community as possible. Under the accord between ESA and NASA for the first Spacelab flight module, Europe was responsible for the development and construction costs of the module while NASA paid for the launch and took responsibility for its operations. In fact Columbia STS-9, the Spacelab-1 mission, was not the first time that elements of Spacelab hardware had flown in space, since experiments mounted on the cargo bay pallet were taken into orbit on the second and third shuttle flights in November 1981 and March 1982 respectively.

FOLLOW-UP MISSION

The first follow-up mission for the Spacelab module did not come until late April 1985, more than 16 months after the inaugural flight. The orbiter Challenger lifted off from Cape Canaveral on 29 April on flight STS 51-B with a seven-man crew aboard, three of them – exceptionally for a space flight – over 50 years old. Even more unusually, it also carried a small ménagerie comprising two squirrel monkeys and 24 rats which were to introduce an element of anarchy into the normally carefully-controlled environment aboard the shuttle. None of the animals actually escaped but mission specialist Bill Thornton, whose unfortunate job it was to look after them, had major problems cleaning and feeding them, since dried food and animal waste shot out of their compartments when feed bars were being changed.

> Something we did not anticipate is that just the slight movement of such material in zero-g is sufficient to send a jet of this stuff out

William Berry, head of the Life Sciences Flight Experiments Division at NASA's Ames Research Center, commented after the flight.[9]

Challenger commander Robert Overmyer was blunter, telling Houston mission control during the flight that 'faeces in the cockpit is not much fun'.[10]

The rats were killed after their return to Earth so that a team of researchers could examine the physiological effects on them of a week in space. The monkeys, having joined the élite NASA astronaut corp, were kept alive, subjected to a series of medical examinations and returned to the life sciences research programme.

Apart from the animal experiments, the second Spacelab flight (confusingly labelled Spacelab 3) successfully showed that it was possible to grow large pure crystals in orbit with potential commercial applications.

The next flight, Spacelab 2, which blasted off on 29 July 1985, was rather different in character, since the pressurised module was left behind on the ground and only the experimental pallets, with instruments housed in an igloo structure, were carried in the cargo bay. The mission, flown by the shuttle Challenger, was essentially devoted to astronomy, astrophysics and solar observation. It encountered a major problem barely five minutes after lift-off, when one of the shuttle's main engines shut down during ascent due to a computer mistakenly diagnosing a non-existent malfunction. A similar problem arose shortly afterwards which could have led to a second, more serious, shutdown, but mission commander Gordon Fullerton manually overrode the computer system. The ascent problems put Challenger into a lower orbit than planned, although still high enough to carry out its scientific programme.

Once in orbit the flight also went far from smoothly, however. Ten years' work by scientists on the ground appeared to be in danger of being wasted when software problems with an instrument pointing system developed by the European Space Agency and Dornier delayed many of the experiments for several days. While a British X-ray telescope worked very well, picking up X-ray emissions from distant galaxies, a Lockheed solar telescope refused initially to activate. It was not until 4 August, the seventh day of the flight, that the crew accidentally kick-started it by making their final attempt to get it to work just after a powerful orbital manoeuvre by the shuttle, which apparently shook it into life. Despite only being able to operate for around one-third of the planned time, the telescope shot nearly 13 000 images of the sun, principally investigating phenomena known as coronal bullets, high-speed jets believed responsible for heating the sun's corona. Software problems abounded with the solar observation experiment, however, compounded by the inability to launch two sounding rockets planned as part of the experiments. One rocket, fired from White Sands, New Mexico, veered off course while the second was never launched due to lightning damage.

Despite the problems, the mission ultimately proved much more successful than controllers had dared hope after the ascent problems and initial difficulties in getting experiments started. The flight was extended by a day to try to recoup some lost time, making for a total of nearly eight days in orbit. Two hundred and thirty miles of magnetic tape were needed to store all the data sent back by Spacelab during the mission.

GERMAN D-1 MISSION

Less than three months later the next Spacelab mission was under way, although no one knew at that stage that it was to be the last before the long interruption to the shuttle programme imposed by the Challenger accident. A record eight astronauts were on board, including three Europeans. Since West Germany had largely financed the flight and was in scientific control of the mission, two of the payload specialists, Ernst Messerschmid and Reinhard Furrer, were West Germans. Both were in their early 40s while a third European, Dutchman Wubbo Ockels, who had been ground back-up on Merbold's 1983 flight, was on board representing the European Space Agency. ESA had a biological research facility, called Biorack, a fluid physics experiment and a sled on board. The sled, which had originally been scheduled to fly on the first Spacelab mission, but which had been held over to simplify the proving flight, was designed to conduct research on the human body in orbit.

The flight was a landmark event, primarily because all scientific activity on board was directed from the German Space Operations Centre (GSOC) at Oberpfaffenhofen, near Munich, instead of from Houston, Texas, transferring primary responsibility for a major part of a NASA flight out of the US for the first time. Apart from the practical advantages, in view of the large number of European experiments on board the flight, the transplantation of control for the scientific part of the programme was a remarkable tribute to the German role in the Spacelab programme.

Lift-off on Challenger came on 30 October 1985 and the crew were in the laboratory within just three and a half hours. As on the first mission, the crew divided into two shifts. In addition to the three European payload specialists there were three US mission specialists on board, Guion Bluford, James Buchli and Bonnie Dunbar. Commander of the Challenger was Hank Hartsfield, at 51 flying his third shuttle mission, while his co-pilot was US Air Force Colonel Steven Nagel.

Furrer, Dunbar and Nagel were assigned to one shift and Messerschmid, Bluford and Buchli the other, with Ockels providing back-up to both shifts and Hartsfield as commander left off the roster. One of the central features of the mission was the space sled, the ESA-designed mobile seat that moved up and down a short track at varying degrees of acceleration with an astronaut strapped in it. The object was to conduct neuro-physiological experiments in weightless conditions, investigating the cause of phenomena such as space sickness. One example of such an experiment was an analysis of the tumbling feeling that people feel when drunk, and which had always been assumed, throughout several years of Earth-bound investigation, to be gravity-related. Investigating the subject in space proved to be a short-cut to understanding the phenomenon. 'With this one experiment it was clear it could not be due to gravity', Ockels said later.[11]

A total of 76 experiments were on board the D-1 laboratory during the flight, and the success rate during the mission was more than 80 per cent, flight scientists said later. In addition to supervising the programmed experiments the astronauts had to conduct a number of unscheduled tasks, including the repair of a mirror heating facility used for material science experiments. The facility was designed to work by creating a vacuum, but a faulty valve prevented it from doing so. Once it was decided to press on with experiments in the facility under the artificial pressure of Spacelab, astronauts under guidance from the ground had to disconnect a sensor, which had been designed to prevent the start-up of the facility until a vacuum had been created. Restricted communications time with the Oberpfaffenhofen control centre, and a series of smoke alarms on board the laboratory, added to the confusion and made ground direction of the repair operation difficult. Messerschmid and Ockels finally succeeded in cutting a cable that disconnected the sensor, enabling some of the experiments involving the facility to continue.

On another occasion the crew had to resort to a saw to mend the fluid physics module on the laboratory, after a valve jammed and a nut was found to be inaccessible to a wrench because it was covered with a plastic cap. The general conclusion of both astronauts and ground scientists was that the mission was a major success, but that this had owed a great deal to the presence of the crew, whose intervention on a number of experiments had prevented them from malfunctioning or being completely lost.

One of the fields in which the flight produced significant scientific results was biology.

On the . . . D-1 mission it was shown without any shadow of a doubt that single cell organisms grow very much faster under microgravity than under ordinary gravity, that the white cells we have in our blood grow rather more slowly inside astronauts, that bacteria become less sensitive to antibiotics and yet proliferate faster. None of these things are explicable on any current hypothesis

one intrigued scientist commented later.[12]

The communications essential to monitoring experiments on the mission were complex, involving the use of two satellites as well as a series of ground stations on both sides of the Atlantic. Voice and data transmissions between the shuttle and the US were bounced off NASA's Tracking and Data Relay Satellite to a ground station at White Sands, New Mexico, on to the Johnson Space Center at Houston, Texas, and across the Atlantic to Oberpfaffenhofen via an Intelsat satellite. In addition direct transmission of some scientific data between the shuttle and ground stations in Europe was possible. Video transmissions from the shuttle reached West Germany several times a day, some live and some recorded, but it was simply too expensive to maintain a live video link between Oberpfaffenhofen and Spacelab throughout the flight. As it was, more than \$1 million was spent on transatlantic communications alone during the mission.

Ockels, reflecting later on ways flight management could be improved in later Spacelab flights, or on a future space station, said an upgrading of communications would be a top priority. 'The first technical change we can expect is an improvement in communications . . . it would be a great help if communications were enhanced', he said, suggesting that one major area for improvement would be to permit astronauts in Spacelab to talk directly by telephone with scientists in their laboratories anywhere in the world, instead of simply at one or two mission control centres.[13]

Challenger returned to Earth on 6 November, landing once more at Edwards Air Force Base, California. It was to be the orbiter's last full flight. Less than three months later, on 28 January 1986, it was blown apart in a fireball barely a minute after lift-off, killing the seven crew on board and halting US manned launches for more than two years.

POST-CHALLENGER DISRUPTION

A further 11 Spacelab related flights were planned during the three years following the D-1 flight, of which five were to include the laboratory itself in either short or long module form, and a further six just with the experimental pallets and protective igloo. The five flights planned involving the manned laboratory were an environmental observation mission on the Atlantis orbiter in August 1986, a life sciences mission on Atlantis in March 1987, a microgravity mission on Columbia in May 1987, a flight involving a Japanese version of Spacelab on Challenger in February 1988 and a life sciences mission, also on Challenger, in July 1988. The pallet flights were mainly to be dedicated to ultra-violet astronomy, solar observation and study of the space environment.

One group of scientists whose work was badly hit by the enforced delays was the team of astronomers and astrophysicists working on the Johns Hopkins ultraviolet telescope, which had been due to fly three pallet missions before the end of 1987 and instead saw flight opportunities slipping back to the end of the decade or early 1990s.

It's beyond the horizon of thinking for any of the young scientists involved ... it's a very big setback. We have to move on to other projects that will come to fruition sooner than that

Arthur Davidsen, principal investigator on the telescope, was quoted as saying.[14]

The interruption of the programme by the Challenger accident, and the delays and cutbacks which inevitably followed, were a severe blow not only to scientists who had frequently worked for four or five years preparing experiments, but also for other potential users of the laboratory, including the US military planners involved in the Strategic Defense Initiative (SDI) who had intended to use it for some laser targetting experiments.

Following the Challenger explosion, and NASA's decision to give priority to military payloads once shuttle flights resumed, it became clear that Spacelab would fly much less frequently than envisaged earlier in the decade.

But aside from the flight value of the laboratory itself, one of its main functions had been to enable European companies to demonstrate their capability of building a manned space laboratory that met NASA's highest technical standards and which gave European astronauts their first taste of life in orbit.

As such it played a key role in paving the way for a much more ambitious programme in the 1990s – the design of a full-scale European manned module as part of an international space station project involving the US, Western Europe and Japan.

9 Remote Sensing, Comets and the Lure of Mars

ORIGINS OF THE SPOT PROGRAMME

The Spot programme was initiated by the French government in 1978 because it wanted the country to have its own Earth observation and photo-reconnaissance technology. Spot itself was designed primarily for civilian purposes while the similar Helios satellite project was for military reconnaissance. The go-ahead for the project came six years after the 1972 launch of the first US Landsat satellite to conduct mineral and crop surveys and monitor pollution.

From the start Spot was an almost exclusively French project under the overall control of the French Space Agency (CNES) and the industrial management of Matra, although Swedish and Belgian companies also became involved. It was conceived as an operational family of up to four satellites, comprising two high-resolution cameras each designed to photograph an area of 60 × 60 kilometres. The craft was to be placed in a polar orbit, meaning that it would circle the globe from north to south, shifting its path slightly on each orbit and over-flying the entire Earth's surface every 26 days.

The pioneer satellite of the series, Spot-1, was launched on an Ariane-1 from Kourou on 21 February 1986, three months later than scheduled due to the failure of the previous Ariane flight, and successfully placed in low Earth orbit at an altitude of 822 kilometres. Within 24 hours of launch it was transmitting its first pictures of Western Europe and Northern Africa to the processing centre at Toulouse. The quality right from the start was very high, with the three spectral bands in the visible and near-infra-red bands on the satellite's cameras capable of producing both monochrome and colour photographs with a resolution of as little as ten metres. More than two months of tests followed to ensure that the cameras and back-up systems were working correctly before the satellite officially became operational on 6 May.

CHERNOBYL – AN EARLY TEST FOR SPOT

By the time it started its commercial life Spot was already being put to its first major test. In late April fire had broken out in one of the four reactors at the Soviet nuclear power station at Chernobyl and information on the extent of the damage both to the plant and the surrounding area was extremely scant. News of the disaster did not surface in the West until radiation began to drift across Scandinavia and was detected by scientists there, prompting an admission from the Soviet authorities that an accident had occurred. On 29 April, three days after the fire broke out, the US Landsat-5 satellite was able to photograph the damaged reactor and two days later, on 1 May, Spot took its first picture of the site. While it did not permit detailed analysis of the extent of the damage to the reactor itself, it clearly showed smoke rising from one part of the plant and the scope of initial damage to the surrounding area. Over the following months Spot was able to monitor the region periodically to provide data on the effect of the radiation fall-out on crops and other vegetation.

In August Spot provided further interesting pictures of the Soviet Union, demonstrating its potential as a reconnaissance satellite despite its official billing as an essentially scientific Earth resources observation vehicle. Spot produced extremely clear photographs of the Soviet Union's secret space shuttle launch and landing facilities at Tyuratam, being prepared for the first flight of the shuttle which eventually took place in late 1988. The Spot photos not only showed the overall shuttle complex but also clearly depicted in considerable detail the assembly buildings, launch pad and landing runway.

Further evidence of Spot's potential military application came with the publication in March 1987 of a series of colour photos taken by the satellite of the Soviet northern fleet's port facilities at Murmansk and Severomorsk. And the 1987 edition of the US Defense Department's report *Soviet Military Power* reproduced Spot pictures showing Chernobyl, a Soviet air base and Soviet anti-missile installations.

Spot suffered a setback in September when one of its two on-board tape recorders failed, making the satellite dependent on the single recorder remaining in operation. The tape-recorders were needed to store images taken by the cameras for later transmission to ground stations at a suitable point in the satellite's orbit. While the failure robbed Spot-1 of some flexibility, it did not seriously impair its operation.

COMMERCIALISATION OF SPOT PICTURES

CNES had set up a commercial company, Spot Image, based in Toulouse, to handle the commercialisation of Spot data and by April 1987, a year after the satellite became commercially operational, Spot Image's two main receiving stations at Toulouse and Kiruna in northern Sweden had registered a total of 270 000 images, of which about 70 000 were sufficiently free of cloud to be commercially usable. In addition receiving stations set up during the year in Canada and the US recorded a further 106 000 images, of which 23 000 were commercially usable. However, processing the images was a time-consuming process, and only a few thousand images had been actually marketed.

Data from the Spot satellite could be used for a variety of different civilian purposes, ranging from cartography to crop surveys and urban planning. Mapping from space was an obvious application of Spot technology, and France's National Geographical Institute (IGN) became involved in the project at an early stage. Since each of Spot's cameras was capable of photographing a 60-kilometre-wide band of the Earth's surface as it passed overhead, the two cameras working together with a small overlap could film a 117-kilometre-wide band when both were pointing vertically downwards. Alternatively the two cameras could take pictures independently of each other either vertically or obliquely, being able to photograph anywhere within a 950-kilometre-wide corridor beneath the satellite. This gave the satellite considerable flexibility in mapping irregularly-shaped areas.

IGN director-general Claude Martinand told a Paris press conference in late 1987 that there was huge scope for the application of Spot data to map both Third-World countries, where accurate maps often did not exist, and more developed nations, where maps rapidly became outdated due to urban development and landscape changes due to farming. Martinand said work had already started with Spot data on mapping several African countries.

> More than half of developing countries do not have maps at a scale of 50,000:1 ... it is clear that it is particularly appropriate throughout the sub-Saharan and Sahel areas because it is easy to acquire images

he said, in a reference to the fact that uninterrupted views from space were easy to achieve over dry areas, whereas cloud cover could be a major problem in equatorial zones which were invisible to Spot for much of the year.[1]

Martinand cited the example of the United Arab Emirates, which had announced an international tender for bids to map the entire country. France had submitted two proposals, one using Spot and one conventional aerial photography. 'The Spot bid was of the order of one-and-a-half to two times cheaper than the offer using aerial photography', Martinand said.

Spot, while following in the footsteps of Landsat, was able to improve considerably on the performance of the US vehicle, offering photos with a resolution of 10 metres compared with 30 metres for Landsat. One of its most striking features was its ability to provide stereoscopic images by photographing the same area from different angles, using the camera set at varying oblique angles during a variety of passes over the same area. This permitted accurate three-dimensional images of the area under study to be generated, which with the aid of computers had wide applications. Urban planners could use the technology to assess the environmental impact of construction projects, for example, while there were also military applications. Computer-generated images accurately reflecting the topography of the landscape could be used to simulate exactly what a fighter pilot would see flying at a specified altitude. Similar technology could also be used to programme ground-hugging cruise missiles, which rely on an accurate three-dimensional map of terrain between their launch site and target area in order to evade radar detection.

Spot data could be used in a variety of ways to assist urban planning and development. It could identify land under forest, agricultural land, rivers and lakes, urbanised areas, construction sites and bare ground, plotting the changing urban landscape.

AGRICULTURAL AND OTHER APPLICATIONS

Its applications for agriculture were perhaps more obvious, however. With its infra-red imaging it was able to differentiate between various types of crops with great accuracy and also detect their stage of development. Because of its swivelling cameras, it could photograph the same area once every two and a half days as it passed overhead, compared with one pass every 26 days for satellites with only vertical-camera capability. This enabled it to plot even fairly rapid changes in crop conditions and detect sudden events such as floods. The data, while of interest to scientists and agronomists, was also of potential interest to supervisory bodies.

With farming heavily subsidised in some form in many countries, governments and supra-national authorities need access to reliable up-to-date information about the state of agriculture in their areas. Jean-Pierre Contzen, director-general of the European Commission's joint research centre, told fellow scientists at a symposium in February 1988 that while the European Community could use remote sensing data in several ways, its potential for enforcing and monitoring the system of subsidies and production quotas known as the Common Agricultural Policy was of paramount importance. 'It's really top of the list', he said, noting that Spot data could be used to distinguish, identify and measure areas planted to certain types of crop, estimate production and help the Commission formulate a management policy for less favoured areas.[2]

It could also be valuable in monitoring subsidies and quotas, checking information supplied by farmers on the ground against satellite intelligence. 'We are using both high and low-resolution satellite data', he said, commenting that the European Community had a $35 billion annual farm budget whose management could benefit from the best-available technological back-up.

The value of Spot images to the European Community is reduced by the large amount of cloud cover over northern Europe for much of the year, however, with only around 20 per cent of images taken over France, West Germany and Britain usable for analysis. The situation is much better further south in Italy, Greece, Spain and Portugal, but comprehensive, Community-wide coverage is patchy. As a result interest is growing in the potential for monitoring crop developments using satellites carrying micro-wave surveillance equipment, to which clouds and other atmospheric disturbances present no barrier. 'To increase the use of remote sensing, let's go to micro-wave as soon as possible', Contzen said.[3]

The Community uses data from remote sensing satellites for a range of other applications apart from agriculture. Environmental protection is high on the list, with satellites capable of monitoring everything from chemical pollution to forest fires, while regional development plans and resource management can also be assisted by satellite data. While Spot Image has been gradually building up a customer base for its material, it has no illusions about ever being able to recoup the cost of the satellite. During the first 18 months of its operation about 10 000 photographs were sold for close to Ffr. 60 million, a performance which enabled officials to predict that the programme would move into operating profit by the mid-1990s. Sales over the first five years of operation were predicted at around 30 000 photographs. A second orbiter, Spot-2, was due to be launched in late 1989, to be followed by Spot-3 in 1992 and the more advanced Spot-4 and -5 in the late 1990s.

The cost of developing the second-generation Spot-4 and -5 satellites, with more sophisticated technology and a longer operational life, was being partly met by the French space agency CNES and partly paid out of the French defence budget, since the military authorities were planning to launch the Helios family of satellites in the 1990s which would be directly developed from Spot technology, providing the country with independent military surveillance capability from space in both the visible and infra-red bands. 'We are sharing part of our development costs with the military', senior CNES official Michel Courtois confirmed to journalists in late 1987.[4]

COMPETITION WITH LANDSAT AND SOYUZKARTA

By the end of the 1980s Spot was the most sophisticated civilian Earth obervation satellite in operation. The US Landsat programme, which had started in 1972 as a purely government programme, was partly transferred into the private sector in 1985 when the National Oceanic and Atmospheric Administration (NOAA) signed an agreement handing management and commercialisation of the Landsat orbiters over to the Earth Observing Satellite Co (Eosat). Its Landsat 4 and 5 satellites were producing large quantities of data, but with a lower resolution than Spot. In addition Landsat was running into difficulties over funds, with the US Congress urging a hold-down in state support for Landsat as part of the more general effort to cut back on the huge US budget deficit. In late 1988 NOAA started talks with CNES of France on the possibility of merging the Landsat and Spot programmes to hold down costs, although this idea ran into some opposition from other US government departments concerned at the wider defence and foreign policy issues raised by such a merger.

In early 1988 Thomas Pyke, NOAA's assistant administrator for Satellite and Information Services, stated that 'the US is definitely committed to achieving a long-term, successful, commercial remote-sensing operation', despite the budget constraints, while adding that 'we are beginning to realise the potential for commercial ownership of remote-sensing systems'.[5]

Apart from Landsat, NOAA was also responsible for running two geostationary GOES remote-sensing satellites and two polar-orbiting Metsat satellites monitoring temperature and humidity, and the organisation was becoming increasingly aware of the commercial potential of weather forecasting and ocean current-monitoring. Revenue from weather forecasting was running at $250 million a year by the late 1980s, with commodity traders, anxious to gauge crop prospects, among the largest clients. Ship owners stood to save large sums of money from information that satellites could provide concerning ice-free shipping lanes, or the strength and direction of ocean currents, if it could be provided in real-time to ships' officers. And hurricane warnings were becoming a routine, but essential, part of the service offered to the wider community which went beyond the scope of daily weather forecasting.

While NOAA struggled to find a secure financial basis for the future operation of Landsat, other Earth observation programmes were going into orbit. Japan's MOS-1 was operational by early 1988, less sophisticated than Spot but providing further evidence of Japan's determination to gain experience in a variety of different space activities on the way towards its goal of becoming a major space power by the end of the century. And the Soviet Union, a veteran of space technology but a newcomer to Western-style commercialisation of the data it could provide, announced in mid-1987 that it was setting up an organisation called Soyuzkarta to market photographs of 15 or 30 metres resolution taken from its Cosmos satellites. Unlike Spot, Cosmos photographs had to be ejected from the satellite in a capsule and physically recovered rather than transmitted from the satellite direct into a waiting computer, so while their resolution was good, their usefulness for rapid computer analysis was more limited. 'They are high-quality photos . . . but I don't think they're competition', Spot Image president Gerard Brachet said.[6]

Soyuzkarta director Anatoliy Nicolaev told Western scientists shortly after the organisation was founded that 'Soyuzkarta has been set up to assist customers in the exchange of scientific data ... the Soviet Union is establishing a permanent system for monitoring natural resources'.[7]

And another Soviet official connected with the programme, Yulian Novikov, said that the Soyuzkarta project offered great potential for monitoring pollution from space, and controlling it, more effectively than by conventional means. 'Customers are not just environmental agencies but also the polluters, such as construction companies', he said.[8]

Amid Earth observation systems already in existence by the late 1980s, and others planned, the field was threatening to become rather crowded. While this promoted healthy competition on the one hand, it also opened the door to wastage of resources and unnecessary duplication of effort by various countries.

In 1988 Eosat president Charles Williams floated the idea of a global organisation to supervise the mushrooming remote-sensing industry and enable it to use its resources most effectively.

> We need some kind of an international council to optimise the limited number of platforms that will be available ... it's still a new industry and it's about to explode. If 50 per cent of all programmes actually fly, it will open up a vast new area of business opportunities around the world

he said.[9]

While the remote-sensing satellites were shaping up to compete with each other, they were also having to compete on price with other, less sophisticated but also potentially cheaper data gathered by more conventional means. Ultimately end-users of Spot, Landsat or other satellite data are concerned to obtain accurate information at a reasonable cost. The trade-off between cost and accuracy becomes quite sensitive when dealing with sophisticated data systems, and data rarely commands a premium simply because it has come from a satellite.

> We're dealing with a computer technology and not a space technology ... the manager doesn't care that the data comes from a satellite, but he does care about the cost

Don Walklet, president of US based Terra-Mar Resource Information Services, observed.[10]

Computer systems have evolved to the point where companies can gain access to computer-treated satellite data with packages priced at less than $50 000, bringing the technology within reach of even small firms.

Commercial demand clearly exists for the data provided by remote-sensing satellites, and the technological progress already made between the first Landsat in the early 1970s and the first-generation Spot satellite launched in 1986 indicates that by the mid-1990s, and certainly the start of the next century, private industry and individuals will have access to a bewildering range of affordable satellite-generated computer data which will make the early attempts at remote-sensing look quaintly primitive.

THE GIOTTO PROJECT

While Spot was a French national scientific programme, the European Space Agency was also developing an interest in scientific satellites and probes. One of the most ambitious projects ever conceived by ESA was the Giotto probe sent to fly close to Halley's comet in early 1986. Packed with scientific instruments and cameras, it was designed to fly as close as possible to the nucleus of the giant comet, sending back to Earth information on its composition and behaviour.

Halley's comet, which orbits the sun on an elliptical path every 76 years, is exceptionally bright and spectacular and so has attracted the interest of astronomers and the wider public for more than a thousand years. It was depicted in the Nuremberg Chronicle in 684 AD, and again in the Bayeux tapestry recording the Norman conquest of England in 1066. Its appearance in 1301 inspired the Italian artist Giotto to incorporate it into his fresco *The Adoration of the Magi*, using it to portray the Star of Bethlehem, and its appearances have coincided with many major historical events, prompting the more astrologically-minded to read great, significance into its periodic returns.

Halley's comet had been extensively studied from the ground over the centuries, and its course had been first plotted by Edmund Halley, who correctly predicted its return in 1758 after studying details of earlier comet appearances. However, its approach to Earth in late 1985 and early 1986 was the first opportunity since the dawn of the space age to study it close-up to gather much more detailed information than had been possible previously.

While the European Space Agency was developing the idea of sending a probe to Halley's comet in the late 1970s, other space agencies around the world were having similar thoughts.

INTERNATIONAL HALLEY WATCH

The Soviet Union was planning to send two probes to the comet, Vega 1 and Vega 2, while Japan's Institute of Space and Astronautical Science (ISAS) was also planning to send two probes, Sakigake and Suisei. In addition, NASA already had a craft in deep space called the International Cometry Explorer (ICE), which had been launched in 1978 as the ISEE-3 (International Sun-Earth Explorer-3) to study solar winds and comets and which was to be steered towards Halley's comet as it passed near Earth.

In 1981 the various space agencies began to realise that there could be considerable advantages in working together on the Halley exploration project, not only to avoid duplicating too many of the experiments but also to use the fleet of six spacecraft, each with their different fly-past times and characteristics, to complement each other. For instance, the value of measurements of the comet taken by the spacecraft closest to it at any given time would be radically enhanced by simultaneous observations by other spacecraft of the environment through which the comet was passing.

An additional advantage of co-operation, as far as the Giotto project team was concerned, was that Giotto was intended to pass far closer to the nucleus of the comet than any of the other spacecraft, and so could benefit from navigational assistance from the other probes to adjust its final approach to put it as close to its target as possible. While Giotto was aiming to fly as close as 600 kilometres to the comet's nucleus, the other probes were designed to pass further out. The two Vega spacecraft were to pass the nucleus at a distance of about 10 000 kilometres, Suisei at 100 000 kilometres, Sakigake close to 10 million kilometres and ICE around 30 million.

The flypasts were designed to cover a period of nearly three weeks, with Vega 1 making the first pass on 6 March 1986, followed by Suisei on March 8, Vega 2 on 9 March, Sakigake on March 11, Giotto on 14 March and ICE on 25 March.

A meeting was held in Padua, Italy in September 1981 when the Inter-Agency Consultative Group was set up to co-ordinate the Halley project, subsequently meeting a couple of times a year in the run-up to the flypast. British Aerospace was awarded the prime contractorship to build the Giotto probe, with SEP of France designing and building the propulsion system. Other companies involved in construction of the probe included Dornier and AEG Telefunken of West Germany, Thomson-CSF of France, Fokker of the Netherlands and Laben of Italy.

GIOTTO LAUNCH AND RENDEZVOUS

It was launched on an Ariane 1 rocket from Kourou on 2 July 1985, just over eight months before the planned rendezous, and during its journey was to join the Soviet and Japanese Halley probes in September in collecting information on solar winds as a back-up to the encounter between NASA's ICE probe and the comet Giacobini-Zinner. A few weeks later scientists controlling the Giotto mission from ESA's space operations centre (ESOC) at Darmstadt, West Germany put the probe through its paces by conducting a full-scale rehearsal of the comet encounter, including a check-out of the multi-colour camera on board supplied by the Max-Planck-Institut in West Germany, which was to take pictures of the comet's nucleus with a resolution of just ten metres from the point of closest approach.

The journey to the comet by the international fleet of spacecraft went smoothly, and on 6 March the Soviet Vega 1 probe made its fly-by, transmitting back a wealth of data despite encountering a dust-jet that knocked out some systems. Vega-2 made its pass three days later 8000 kilometres from the nucleus, around 900 kilometres closer than Vega-1. Despite some damage to both spacecraft during the encounter, Soviet scientists were jubilant at the results, contributing to the mounting excitement at ESOC as the days ticked away to the Giotto flypast. Two days before the rendezvous European scientists had a moment of alarm when communications with the probe were apparently lost. After years of preparations and an eight-month journey through space, it seemed impossible that a major breakdown could occur when the comet was so tantalisingly close. In fact the problem turned out to have a banal cause which was easily remedied. A workman in Australia with a mechanical digger had accidently severed a cable linking two ground stations involved in tracking the probe, blacking out communications.

Giotto's final approach to the comet was given considerable publicity in Europe, with some television stations staying open late into the night to take live transmission of the pictures as they arrived in Darmstadt. To the disappointment of some television viewers, who had expected optical film of the comet, the pictures were artificial-colour computer images which required considerable explanation for the layman to interpret. But to the scientists at Darmstadt they represented a goldmine of information which would take years to analyse and study. As the probe sped towards the comet there were considerable fears that collision with dust particles could severely damage or even totally knock out its scientific instruments. While Giotto was not expected to survive its encounter with the giant comet, it was designed to keep transmitting during most of its passage through the area closest to the nucleus. In the event signals remained very clear until the moment of closest approach, when Giotto was travelling at a relative speed of around 245 000 kilometres per hour. It passed within 590 kilometres of the nucleus at 01.03 local time in Darmstadt (00.03 GMT) on the night of 13/14 March.

Two seconds before closest approach dust particles disturbed the orientation of the probe's antennae, interrupting reception of most signals for around half an hour until communications could be restored. While the interruption was frustrating it had been largely anticipated and came sufficiently late in the encounter to ensure that mission scientists had more than enough data to work on.

The mission was accomplished beyond all expectations since we survived the encounter, and the spacecraft continued to provide data even after the point of closest approach

said Roger Bonnet, ESA's director of scientific programmes.[11]

While Giotto signals were lost as it flew out from the comet, the two Vega spacecraft had continued to transmit during the outward-bound leg, providing images from both sides of the nucleus. Taken together, the data from the three probes provided a mass of information about the physical characteristics of the comet's nucleus, described by scientists as roughly peanut-shaped and apparently with crater-type features. Around 1000 astronomers grouped within the International Halley Watch were involved in analysing the data from the fleet of spacecraft. However, Giotto data was being primarily handled by a European team while Vega data was being studied by Soviet scientists in collaboration with Hungarian, French and East German researchers who had also had experiments on board the probes. A conference was held in Heidelberg, West Germany in November 1986 to assess work done on the Halley data, and over the next couple of years International Halley Watch worked on storing the huge quantity of information and analysis on computer discs.

In November of that year the European Space Agency decided that Giotto was probably in sufficiently good condition to consider planning a fly-past of a second comet, Grigg-Skjellerup, in July 1992, although further financial and technical studies would be needed before a decision on the feasibility of the mission could be reached.

THE PHOBOS PROJECT

On 7 July 1988 the Soviet Union launched the Phobos-1 probe on a seven-month odyssey to Mars, followed five days later by its twin, Phobos-2. The probes followed a long series of missions to Mars, a planet which had held particular fascination for humanity down the centuries due to its relative closeness, the eery beauty of its distinctive red colour and fanciful speculation that the canal formations on its surface might have been constructed at some time in the past by alien beings. The US Mariner probes in the 1960s observed Mars, photographing it up close for the first time and in 1971 Mariner 9 went into orbit around the planet and studied its atmosphere. Between 1971 and 1973 the Soviet Union sent six Mars probes to the planet, and in 1975 the US sent Viking 1 and 2 to photograph the planet's moons, Phobos and Deimos, and to place landing vehicles for the first time on the planet's surface. The two vehicles, each weighing one tonne, were equipped with two television cameras, soil analysis equipment, seismometers and meteorological instruments which sent back a mass of data to Earth but, sadly for the romantics, failed to find evidence of any little green men past or present.

The Phobos probes launched in mid-1988 were the first attempt at large-scale international collaboration on a scientific space programme since the immensely successful Giotto and Vega probes, and expectations were therefore high that the mission would accomplish similar results. While the Soviet Union was in charge of the programme, a dozen other European countries from both sides of the iron curtain were involved – East and West Germany, Austria, Bulgaria, Czechoslovakia, Finland, France, Hungary, Ireland, Poland, Sweden and Switzerland, as well as the European Space Agency. Altogether about 30 experiments were aboard the probes, which were designed to go into orbit around Mars in late January 1989, manoeuvre themselves into an orbit close to that of Phobos and then swoop down to within 50 metres of the moon's surface for up to 20 minutes before each releasing two small experimental pods onto the surface.

One pod, called the DAS, would remain stationary for about a year, transmitting data back to Earth, while the second, more picturesquely referred to as the Frog, would be equipped with mechanical legs allowing it to move around the surface in 20-metre hops, exploring its immediate environment. The instruments on board the DAS included a spectrometer to study the composition of the moon's surface, a seismograph to investigate the internal structure of Phobos, cameras to observe the microstructure of the surface directly around the pod and a range of instruments to measure the temperature and the physical and mechanical properties of the surface. The Frog, while having some similar instrumentation to the DAS, was equipped with an accelerometer to measure the strength of acceleration caused by its impact while moving around the surface, a spectrometer to study the chemical composition of the surface and a magnetometer to study the local magnetic field.

The cost of the whole programme was close to half a billion dollars, a huge amount of money to spend on investigating an apparently barren potato-shaped lump of rock circling an uninhabited planet. What was the point of such a large investment?

THE LURE OF MARS

From the scientific point of view the principal interest of Phobos is that it appears to be one of the most primitive bodies in the solar system, a remnant from the period before planets existed which somehow became trapped in a Martian orbit and which could yield valuable clues to the way the planets were formed. But the exploration of Mars and its moons also had a symbolic value.

> Beyond the intrinsic scientific interest which such a body, containing a goldmine of planetary phenomena, represents for comparative planetology – geological activity affected by wind, water ice, an active atmosphere, climatic variations – Mars has become, for the major space agencies, an important step on the road to the conquest of space

the French space agency CNES said in a document marking the venture.[12]

The Soviet Union also saw the study of Phobos as one stage in the preparation for a full-scale joint manned mission to Mars with the United States early in the next century, giving it a critical role not only in the international space science programme but also in the longer-term manned exploration of space.

LOSS OF RADIO CONTACT

Therefore it came as a bitter blow when radio contact was lost with Phobos-1 on 2 September, less than two months after its launch when it was just 17 million kilometres from Earth. An error by a ground engineer was responsible for severing the link, which Soviet scientists spent two months trying vainly to restore before finally giving up in early November. The accident, doubly frustrating because of its trivial cause and avoidability after so much pain-staking preparation for the mission, left just one probe, Phobos-2, operational and still heading for Mars.

Despite a report in the *Houston Chronicle* in late December that Phobos-2 was also experiencing severe mechanical problems, the probe swung into orbit round the red planet right on schedule at 12.55 GMT on Sunday 29 January 1989, six-and-a-half months after launch.

It immediately began gathering data on the Martian atmosphere and during the following eight weeks surveyed the ionosphere and magnetosphere, conducting a mineralogical and chemical survey of the planet and transmitting data to help improve existing maps of its surface. It also provided information on solar gamma flashes.

Press interest in the epic voyage picked up as the time neared for the flypast of Phobos, and the landing of pods on its surface. 'The invasion of Mars starts at Phobos' trumpeted the Italian daily *La Repubblica* on 28 March, splashing photos of Phobos across a full-page feature article.

The irregular, hump-backed outline of Phobos, the smallest of Mars' two moons and the first link in the long and difficult chain which will take man to the surface of the Red Planet, appeared on Easter Sunday on the video screens at the Moscow control centre ... it is here that lies a possible jumping-off point for the conquest of Mars

the paper wrote enthusiastically.[13]

Ironically for *La Repubblica*, and disastrously for the scientists who had staked years of work on the project, the Phobos-2 mission was facing disaster even as that article was being written. For on Easter Monday, 27 March, as scientists were trying to rotate the probe so that it could take more photographs of the Martian moon, radio contact was abruptly lost. It was a bitter disappointment after the huge scientific effort invested in the project, and the hundreds of millions of dollars spent.

Despite the blow further Mars flights were planned for the 1990s as preparations continued for a possible manned mission. The US was scheduling a Mars orbiter mission (MOM) for around 1992 to study its geophysics and climate, while both the US and Soviet Union were planning unmanned flights by the end of the century to retrieve soil samples and bring them back to Earth.

METEOSAT AND MOP WEATHER SATELLITES

While the Phobos mission was rather unsuccessfully trying to push back the frontiers of space exploration, work was continuing in Europe on a much more down-to-Earth programme which nevertheless had a considerably greater daily impact on millions of people – the Meteosat weather satellite system. Meteosat was Europe's contribution to the World Weather Watch global observation system run by the World Meteorological Organisation (WMO), providing meteorological coverage of the whole of the Earth's surface through a chain of five geo-stationary satellites, two US, one Japanese, one Indian and one European. These were stationed over the Atlantic, Pacific and Indian oceans and were complemented by US and Soviet polar-orbiting satellites.

> Observations from space are an essential component of the global observing system ... meteorological observation is the most highly developed area of the more general field of remote sensing, the benefits of which are only just beginning to be recognised and realised

the British Meteorological Office commented in early 1987.[14]

Meteosat-1 had been launched in November 1977 and Meteosat-2 in June 1981, the two orbiters complementing each other since the imaging equipment on one and the data gathering instruments on the other failed. A third satellite, Meteosat-3, was placed in orbit in June 1988 aboard the first Ariane-4 rocket to avoid the possibility of the service being interrupted should the two original satellites cease to function before more modern replacements could be launched.

In fact the first of the new generation of weather satellites, MOP 1, was launched less than nine months later, in early March 1989.

MOP 1 was the first of a three-series family of satellites designed to be operated by Eumetsat, which had been founded in June 1986 to run the European meteorological satellite programme on behalf of 16 European countries. With a budget of around 530 million European Currency Units (ECUs) at 1989 prices, the bill was largely met by four countries, West Germany paying 26.39 per cent, France 25.60 per cent, Britain 16.76 per cent and Italy 12.00 per cent.

MOP 2 was due to be launched in 1990 and MOP 3 between 1992 and

1994, with the older Meteosat-3 remaining in orbit as a back-up vehicle. The new satellites weighed just 680 kilos on launch, including their apogee motor, and each had four fully independent imaging channels and two communication channels. The satellites were designed and built by France's Aérospatiale, with co-contractors including MBB and ANT of West Germany, Matra and SEP of France and Selenia Spazio of Italy.

With the Meteosat programme well into its second decade of operation, and the original two satellites having at least partly outlived their three-year design life, space officials have expressed considerable satisfaction with its contribution to weather forecasting and are in no doubt of the need to develop the programme further. 'Meteosat has been, and still is, one of the major achievements of the European Space Agency and of Europe', P. Goldsmith, director of ESA's Earth Observation and Microgravity Programme, told scientists in Darmstadt marking the tenth anniversary of the programme.[15]

And Meteosat official Brian Mason, speaking on the occasion of the MOP 1 launch, commented that

> meteorology today needs a global vision of what is happening on the planet, both at altitude and on the ground, but above all in the two-thirds of the Earth which are virtually uninhabited, consisting of the deserts, the great forests, the poles and the oceans. Only an integrated network of space observatories can provide this.[16]

The MOP series of weather satellites, while intended to provide an operational service until close to the end of the century, was itself a technological bridge to a more advanced and complex generation of weather-forecasting space-based observatories being planned for the 1990s.

The European Space Agency's ERS-1 and ERS-2 satellite project was already at an advanced stage of development when MOP-1 was launched, with ERS-1 scheduled to go into orbit in 1990 and the identical ERS-2 possibly slated to follow it in 1993. The satellites were designed as remote-sensing orbiters to monitor the oceans and polar ice-caps, measuring such phenomena as wave heights, wind speeds, sea-surface temperatures and cloud behaviour. The satellites were therefore designed with a multi-role mission, partly pure and partly applied science, but clearly had direct applications for meteorology.

And further in the future a polar orbiting platform planned by the European Space Agency as part of its contribution to a US-led international space station project in the late 1990s would probably also be equipped with instruments designed for weather monitoring, supplementing data coming from meteorological satellites already in orbit.

10 The Telecommunication Satellite Revolution

The groundwork laid during the 1970s for satellite telecommunications in Europe was rapidly built upon during the following decade. The backbone of the system was provided by Eutelsat, the 26-nation organisation which by the end of the 1980s had four operational satellites in orbit, twice the number it had originally envisaged. The reason for the extra capacity was partly the surge of interest from television companies in the possibilities offered by satellites, initially for news and sports coverage but later for a wide range of other transmissions.

> TV was a market which nobody had anticipated in 1977/78 when Eutelsat was formed. Because of this Eutelsat had planned initially to launch no more than two satellites of the first generation, one providing services and the other one as a spare in orbit. Now this big boom in TV, and the introduction of the business services ... required a complete change in the operational planning of Eutelsat

Andrea Caruso, Eutelsat Director-General, said in an interview shortly before his resignation.[1]

EUTELSAT AND NATIONAL NETWORKS

The first of the Eutelsat satellites, ECS-1, was launched in June 1983 by the European Space Agency aboard an Ariane, successfully as it turned out despite the rocket's uninspiring track record at that time of two failures in just five flights. Its sister, ECS-2, went into orbit in August the following year, but the third flight model, ECS-3, was destroyed when Ariane V15 blew up in September 1985. The replacement third satellite, ECS-4, had to sit out the long delay caused by the subsequent failure of Ariane V18, but was finally orbited in September 1987, while a fourth vehicle, ECS-5, joined it in space the following July. The satellites, weighing just over one tonne at launch, were designed to have an operational life of seven years and were each equipped with ten transponders for simultaneous use and between two and four back-ups. ECS-5, once launched, was manoeuvred into ECS-1's slot at 13 degrees East and television and radio channels which had been using the older satellite were transferred onto the new one, freeing up ECS-1 to be moved to a new position.

Around 8000 telephone circuits, directly linking international centres, were available by the end of 1988 on the Eutelsat system, while it also carried more than a dozen television channels and ten mono and stereo radio channels for cable transmission or direct reception through domestic satellite dishes.

Alongside the Eutelsat network France had orbited its own system of communications satellites, Télécom 1A in August 1984 and Télécom 1B in May the following year. The satellites were built by two French space electronics companies, Matra as prime contractor and Alcatel Espace, which was responsible for the telecommunications payload. The satellites were designed to provide business communications and television links within France and neighbouring countries as well as telephone and television links between France and its overseas departments and territories ranging from French Guiana, Martinique and Guadeloupe in the Caribbean to the Indian Ocean island of Réunion. The satellites also provided links for French government communications, notably supplying French military communications for ship- and land-based mobile receiving stations through the Syracuse system.

The Télécom satellite programme was an illustration both of France's political determination to be the leading European space power with its own civilian telecommunications system, and also its military need as a nuclear power with a self-imposed global role for a sophisticated, secure, totally independent satellite communication system, not least to ensure round-the-clock contact between the French president and military chiefs in Paris and the submarine-based nuclear strike force patrolling the oceans.

The Syracuse system, an acronym for 'Système de Radio-Communications Utilisant un Satellite' (satellite radio communication system), was developed jointly by Alcatel Espace and Thomson-CSF and used a wide variety of ground receivers. Fixed stations in France had large eight-metre diameter antennae while the army used mobile three-metre dishes mounted on trucks, or smaller 1.3 metre dishes. Naval ships were equipped with 1.5 metre antennae. The whole Syracuse system was compatible with the RITA battlefield communications system developed by Thomson-CSF for the French military administration, and which was subsequently sold to the US army in one of France's biggest high-technology military export deals of the 1980s.　　　　　　　　　　　　　　　　　　　　　　　　—

The system basically performed well, although sporadic problems occurred with the Syracuse payload on Télécom 1A, which occasionally switched itself off due to build-ups of excessive electrostatic charges. The problem had been detected in its early months of operation, and some modifications were made on its sister satellite prior to launch to prevent a repetition of the same phenomenon. As a result it was decided eventually to transfer military communications to Télécom 1B. But before this was able to take place Télécom 1B suffered a sudden and catastrophic failure of both its main and back-up attitude control systems on 15 January 1988, just over half-way through its seven-year design life. The failure of one component caused a chain reaction of failures through the satellite, leaving it orbiting uselessly.

TELECOM 1B FAILURE

As Claude Goumy, director-general of Matra's defence and space division, commented a few weeks after the accident:

> Télécom 1B had the most extraordinary piece of bad luck. Most of the time, if you lose a satellite, it is at the start of its life. If it works, generally it works for a long time, or you see a gradual deterioration ... but you have some capacity remaining. The total loss of a satellite after three years is really a one in a million occurrence.[2]

The failure of the satellite so early in its life left the French military, as well as a host of television and radio stations, perilously dependent on just one operational satellite, with no back-up in orbit, and created capacity strains which put in jeopardy some of the peripheral services of the satellite. But if the Télécom 1B accident had been a piece of unexpected bad luck, this was offset by the reciprocal good fortune that the third satellite in the series, Télécom 1C, had been routinely scheduled for launch on the very next Ariane flight. This launch, originally programmed for the end of December prior to the Télécom 1B problem, had been postponed until early March because of concern over an Ariane engine, and now became a matter of urgency.

Matra engineers took advantage of the fact that the satellite was still on the ground to modify it slightly to prevent the possibility of a similar malfunction to the stricken Télécom 1B.

> We have taken every step to ensure that this catastrophic failure can in no way repeat itself ... this required some minor modifications on the Télécom 1C instruments and also on operating procedures. Now we can say that Télécom 1C will work very well for at least seven years

Goumy said.[3]

The launch of Télécom 1C took place on 11 March, and everything went according to plan, with the satellite easing into a perfect orbit. 'I think it's an absolutely fantastic success', beamed a relieved Marcel Roulet, director-general of France Télécom as the satellite spun clear of the rocket.[4]

Officials of the organisation said it would not only permit it to guarantee continued service to its existing clients, such as private French television stations La Cinq and M6, but also to offer new cut-price contracts to companies interested in renting spare capacity, on the understanding that their service would be cut off first in the event of any technical failure.

Apart from bringing relief to the French telecommunications industry, the same Ariane flight coincidentally marked the first launch of a US telecommunications satellite since the Challenger accident of January 1986. The rocket blasted the Spacenet IIIR satellite, owned by GTE Spacenet, into an orbit 87 degrees West, serving the entire continental United States and complementing a series of satellites already in orbit. 'It's been two years since the US had a successful launch for a communications satellite. This is a very important revitalisation of the communications satellite programme in the US', GTE Spacenet president Gerry Waylan commented in Kourou after the launch.[5]

As a result there were celebrations on both sides of the Atlantic when the Ariane flight passed off without incident.

BRITAIN AND SKYNET

While France relied on the Télécom satellite system for its military communications, other European countries used a system of North Atlantic Treaty Organisation (NATO) satellites. But Britain had also designed and built a dedicated military satellite system, Skynet 4, which by the end of 1988 was ready for launch. The system, comprising three satellites, was constructed for the British Ministry of Defence by British Aerospace as prime contractor, with Marconi Space Systems as subcontractor for the communications payload.

It was designed as a follow-on to a sequence of earlier satellites, starting with Skynet 1A, the first British military communications satellite, which was launched in November 1969 and which performed well in orbit for two years, a respectable life-span given the state of technology at the time. The second Skynet 1 satellite and the first vehicle of the subsequent generation, Skynet 2A, were both destroyed by launcher failures but Skynet 2B, which went into orbit in 1974, was still functioning more than a decade later. Despite a fault which robbed ground controllers of the ability to steer the satellite in its orbit, it continued to provide sporadic communications links into the mid-1980s and was used during the Falklands campaign of 1982 for communications between London and the task force in the South Atlantic.

Accords with the US and NATO giving Britain access to their military communications satellites, combined with a cut-back in Britain's global role following the government's decision to terminate permanent garrisons east of Suez, led to the cancellation of the planned Skynet 3 series in 1975, but pressures on capacity caused by mounting traffic in the 1980s subsequently led to the commissioning of the Skynet 4 series. The first Skynet 4 had been scheduled for launch on the shuttle in May 1986, to be accompanied into space by a British astronaut, but that mission was cancelled following the Challenger accident, and the first Skynet 4 satellite,

4B, was ultimately launched on an Ariane in December 1988 after yet more delays caused by the 16-month Ariane launch interruption.

> The Ministry of Defence is increasingly dependent upon space communications . . . the tragic loss of Challenger Shuttle in 1986 and subsequently problems with Ariane have set that programme back quite severely . . . meanwhile we are coping, albeit with some difficulty

Air Vice-Marshal E. H. Macey, Assistant Chief of the British Defence Staff for policy and nuclear matters, told a Parliamentary space committee in October 1987.[6]

Skynet 4B was the first purely military satellite ever launched by Arianespace, which had been set up primarily to market launch space to civilian operators. It was designed to give maximum communications flexibility to the British armed forces, providing a variety of options ranging from highly-focussed spot coverage to a beam spanning the entire hemisphere centred on the Greenwich meridian. This provided communications capability over the whole of the NATO area, the Mediterranean, the North and South Atlantic, the Middle East, Gulf and parts of the Indian Ocean. It had five transponders on board, a launch weight of 1.4 tonnes and a mass in orbit of 790 kilos. The ground control centre, built by Marconi, was sited at RAF Oakhanger.

Not only was the Skynet 4 programme a revitalisation for British military communications, but also a shot in the arm for the country's space industry, which had been running short of official funding.

> The re-emergence of the national military satellite programme Skynet 4 has enabled the UK to secure orders for the supply of two Skynet 4 type satellites to NATO to meet Alliance requirements through the 1990s. The success of the Ministry of Defence bid – in competition with the US – has provided a significant boost for the UK space industry

the Ministry of Defence said in a memorandum to Parliament.[7]

The NATO order, worth close to $280 million, of which nearly half went to British Aerospace and Marconi Space Systems, was announced in January 1987 and marked the first time that the Alliance had awarded a satellite contract to a non-American supplier.

The two NATO satellites, called NATO 4, were for launch in the early 1990s to provide bug-proof communications for both military and diplomatic traffic across the Alliance until the end of the century. British Aerospace and Marconi won the bid against competition from General Electric of the US, and the first of the NATO satellites was slotted for launch on the shuttle in 1990. The NATO 4 satellites were virtually identical to Skynet 4, and similarly a derivation of the original Orbital Test Satellite (OTS) launched by the European Space Agency in 1978, which had proved such a success.

As for the remaining British satellites, Skynet 4A was due to ride into space aboard a Martin Marietta Titan 4 rocket in late 1989 while an Ariane was due to place Skynet 4C in orbit in mid-1990.

WEST GERMAN AND ITALIAN SYSTEMS

Two other European countries which had active telecommunications satellite projects were West Germany and Italy. The West German Bundespost (Post Office) had commissioned two DFS Kopernikus satellites from a consortium led by the two West German space electronics groups MBB-ERNO and ANT. The satellites, with a launch weight of 1.4 tonnes, were designed with 11 transponders each, to provide multi-purpose telecommunications links within West Germany and West Berlin as well as neighbouring countries. DFS-1 was launched on Ariane in mid-1989 and DFS-2 scheduled to follow in early 1990. Italsat, ordered under Italy's national space plan from its leading space electronics supplier Selenia Spazio, was slightly larger, with a launch weight of 1.75 tonnes, and was due to go into orbit in the early 1990s.

Apart from the national military and civilian satellite communication systems and the Eutelsat network, European countries were also involved in two global space-based telecommunications organisations, Intelsat and Inmarsat.

INTELSAT AND INMARSAT

The International Telecommunications Satellite Organisation (Intelsat), with its headquarters in Washington DC, was set up in 1964 with 11 member countries with the goal of providing global telecommunications links. Its first satellite was Early Bird, weighing just 40 kilos and launched in 1965, and there followed a whole series of satellites, each larger and more sophisticated than their predecessors. The organisation ran into major problems with its third-generation Intelsat III series in the late 1960s, when four of the six satellites either never reached orbit or failed soon afterwards. But it pressed ahead with the fourth and fifth generations, having more than 20 satellites in orbit by the late 1980s. There were 15 flight models built of the Intelsat V series, which had a launch weight including fuel of nearly two tonnes and an orbital mass of 1.1 tonnes, while the succeeding Intelsat VI series, scheduled to start becoming operational in late 1989, was more than twice the size. Ford Aerospace signed a contract in 1988 to build five Intelsat VII satellites, with options for more, for launch from 1992 onwards, assuring continuity of service through to the end of the century. French group Alcatel Espace had a significant subcontract on the programme to provide much of the telecommunications payload, which would provide capacity for up to 190 000 simultaneous telephone calls and three television channels on each satellite, compared with the 15 000 calls and two television channels on the Intelsat V series. By the end of the 1980s Intelsat had over 100 member nations and served around 170 countries worldwide.

While Intelsat was primarily concerned with serving clients with fixed receiving stations, the International Maritime Satellite Organisation (Inmarsat), based in London, specialised in mobile satellite communications. Inmarsat was set up in 1979 and formally began operations in 1982, with a membership which had grown to more than 50 countries by the end of the 1980s. It functioned as an international co-operative, using a network of nine geostationary satellites spread around the globe which gave virtually worldwide coverage, except for the extreme polar regions and a vertical strip running through part of the western US and southeastern Pacific. Two of its satellites, Marecs A and B2, were leased from the European space agency and had been in orbit since 1981 and 1984 respectively, Marecs B having been lost in September 1982 due to an Ariane failure. Three further satellites were taken over from the Marisat system, which had been operated until the early 1980s by the US group Comsat General, while Inmarsat also leased space aboard four Intelsat V orbiters. Three of the satellites were stationed over the Atlantic, Indian and Pacific oceans providing the main communications network, while the other six functioned as spares.

Inmarsat had started out developing a communications network primarily linking ships at sea to coastal stations, providing telephone, telex and data links. Natural offshoots of these services included facilities to co-ordinate search and rescue operations, transmit messages to a whole fleet of ships simultaneously and track automatically the position of ships in the oceans. Services that developed from these facilities ranged from the transmission of facsimiles of weather charts and other technical data to ships to providing such items as newspapers for passenger liners. For a period the Paris-based *International Herald Tribune* used the service, producing a special edition for passengers on Britain's QE2, edited at its head office and sent out over the Atlantic by Inmarsat.

Apart from ship-to-ship and ship-to-shore communications, however, Inmarsat also began to branch out into a related area which offered huge market potential, that of communications between aircraft and the ground. For a trial period between October 1987 and February 1988 the world's first credit-card telephone calls from a passenger aircraft were made from a Japan Air Lines (JAL) Boeing 747 flying from Tokyo across the Pacific to the US West Coast. The field tests were intended to precede the installation of such facilities on many of the world's airlines. Apart from such passenger services, Inmarsat was also beginning to develop services applicable to airline operations and cockpit communications.

Another growing area of activity was that of land mobile communications, specifically as they applied to the trucking industry. Satellites were being developed which could not only relay data to receiver terminals on board large trucks giving such information as loads to be collected or dropped, or the state of the road ahead, but could also permanently track the progress of the truck and report its position back to company headquarters. While some truck drivers were wary of a technology which appeared to have rather sinister Big Brother aspects to it, trucking firms were becoming aware of the potential that such systems offered for improving management efficiency and ultimately cutting costs. While land mobile systems were becoming operational in the US, they were still relatively unknown in Europe, where their market potential was expected to increase with the opening up of European Community frontiers as tariff barriers between Community states disappeared.

Inmarsat's new generation of satellites, the Inmarsat II series, was under construction in the late 1980s by an international consortium headed by British Aerospace, and due for launch by the early 1990s. With launch weights ranging from 1.1 to 1.3 tonnes, they would each offer 250 two-way voice circuits, five times the capacity of the earlier generation. With the arrival in orbit of the new generation, Inmarsat hoped to fill in gaps in its existing world-wide coverage and make its satellites available on a truly global basis.

11 Satellite Television

The era of satellite television in Europe was slow to get off the ground, to the frustration of some who advocated greater choice for viewers and the relief of others who feared that an explosion in television air-time would lead to a drastic decline in the quality of programmes. Throughout the late 1970s and 1980s, as television satellite projects gradually gathered momentum in Europe, legislators and special interest groups looked to the US as a guide to the way that European television might develop. The huge quantity, and often dubious quality, of channels available in the US through a mixture of cable, satellite and Hertzian wave broadcasting sparked a lively debate back on the old continent, with cliché phrases such as 'wall-to-wall Dallas' – evoking the spectre of 24-hour American soap in European living rooms – coming to dominate dinner party conversation from Maida Vale to Malta.

OBSTACLES TO SATELLITE TV

There were innumerable barriers to the establishment of satellite broadcast services, some technological, some cultural, some political or bureaucratic and some plain financial. On the technological level the satellites had to be designed, built and – still the most hazardous part in the mid-1980s – launched. On the cultural level the West European audience of 300 million people spoke a dozen separate languages and had very divergent tastes and traditions, making it radically different from the US market which, despite different language groups and cultures, was much more homogeneous.

Politically some countries saw the launch of a television satellite as a symbol of national prestige, rather like owning an airline, or as a means to expand their cultural influence across borders. Bureaucratically there was a maze of different regulations in the various different countries governing advertising and broadcasting rights to cultural and sporting events and films, as well as such issues as the screening of sex and violence. Broadcasting laws had grown up haphazardly over the previous 30 years, and no overall guidelines existed. And financially, despite the attractions of the satellite television business on paper, there was little hard evidence that it could be turned into a serious profit-making exercise, or that viewers would be prepared to pay relatively large sums of money for the equipment needed to receive the broadcasts.

The number of people who will be persuaded to buy new domestic equipment because of the technological beauty of satellites will be

110

limited indeed. The battle will be won or lost on the quality of the programmes: the new broadcasters will need to produce programmes for which the consumer is prepared to pay, or they need to be attractive enough to provide audiences too interesting for advertisers to ignore

the *Financial Times* wrote.[1]

Prior to the arrival of satellite television, most Western European countries had between three and four national channels available to viewers. State control was gradually giving way to private ownership on some channels, and near borders viewers were able to pick up a variety of broadcasts from neighbouring countries, making for wider choice. But coverage was patchy, with the result that in some areas, particularly mountainous regions, viewers had difficulty even picking up all of the state channels, while in parts of the Netherlands viewers could tune in to as many as 15 channels – Dutch, Belgian, German, French and British.

During the 1980s several satellite channels sprang up, transmitted on existing low-power telecommunications satellites like the Eutelsat-1 ECS series. Entertainment channels such as Sky, owned by Australian-born media mogul Rupert Murdoch, Superchannel, owned by a group of British companies including Virgin, and the Lifestyle and Screensport channels controlled by British bookstore chain W. H. Smith, were all English-language products. German-language viewers could see Sat 1 and Filmnet, French-language viewers TV 5 (run by a consortium of French, Belgian and Swiss channels) and Italian viewers outside Italy a satellite relay of the main state channel Rai Uno. Worldnet carried English-language press conferences from Washington and other capitals to journalists across the continent, while Ted Turner's 24-hour Atlanta-based news channel Cable Network News (CNN) kept travellers and news addicts in smart hotels in Paris, Rome and Munich up to date with US affairs not normally covered on European news broadcasts, such as the latest baseball scandal, minutiae of power politics in Washington or the weather outlook for the Great Lakes.

However, while it was technically possible to receive many of these channels in parts of France, West Germany and the Benelux countries at the centre of the Eutelsat beam with a 90-centimetre receiving dish, in practice very few people did this and viewing in the mid-to-late 1980s was almost exclusively through cable, either in the relatively small number of homes connected to cable distributors, or more usually in hotels. It was only towards the end of the decade that the viewing public began to wake up to the radical change in the television landscape which was about to hit it with the advent of direct-broadcast satellite (DBS) systems.

While DBS operators expressed confidence that the public would be anxious to buy dishes to receive new channels directly in their homes, others in the industry were more sceptical, seeing a growth in distribution of satellite broadcasts through cable networks as a more likely result.

By 1998 we think nearly 30 million homes in Western Europe will receive satellite television but only about one-and-a-half to two million will be receiving direct, the rest will be on cable which we see growing at nine per cent a year

Patrick Whitten, managing director of British communications consultants CIT Research, predicted in mid-1988.[2]

THE FRANCO-GERMAN INITIATIVE

The first initiative taken in Europe to get a DBS system off the ground was in April 1980, when France and West Germany gave the go-ahead to their industries to develop jointly a series of four identical satellites, the French TDF-1 and 2 and the German TV-SAT 1 and 2. West German industry had 54 per cent of the work on the project and the French 46 per cent, with the contract being awarded to a Munich-based consortium called Eurosatellite GmbH. This had been set up in 1978 by six European aerospace and electronics companies, AEG, ANT Nachrichtentechnik and MBB-ERNO of West Germany, Aérospatiale and Alcatel-Espace of France and ETCA of Belgium, to develop broadcasting satellites. Project management for the TV-SAT 1 and 2 and TDF-1 programmes was at MBB's facility at Ottobrunn, in the Munich suburbs, while the TDF-2 project was managed from Aérospatiale's satellite facility at Cannes. Government responsibility for the programmes was handled in West Germany through the Federal research and technology ministry (BMFT), the aerospace research establishment DFVLR and the Post Office, while in France responsiblity was handled by the space agency CNES and the broadcasting authority Télédiffusion de France (TDF).

Total development and construction costs of the joint programme were estimated at 900 million marks and launch of TV-SAT 1 and TDF-1 were originally scheduled for 1984, with the other two satellites to follow later in the decade. The satellites were designed partly as an experimental system, proving the technology and paving the way for a later generation of DBS satellites, but if successful were also envisaged as operational satellites charging commercial rents to broadcasters to transmit programmes.

Each satellite was designed to have five channels, four of which could be used simultaneously and the fifth kept in reserve in case of failure. Each channel could handle one television programme or 16 radio programmes, permitting each satellite to broadcast, for example, either four television channels or three television and 16 radio channels simultaneously. Viewers could receive programmes with a 50-centimetre dish in the central area of the satellite footprint, and a 75- or 90-centimetre dish in more outlying areas of Europe. The French satellites were to be centred, naturally enough, on France, but would also cover an area stretching from central Spain and northern Portugal to Scotland, Denmark, Italy and Yugoslavia. The TV-SAT orbiters, centred on West Germany, would also be picked up in East Germany and much of Eastern Europe and southern Scandinavia, as well as eastern France and northern Italy. They were designed to last at least seven years, and possibly as much as nine if placed in a particularly accurate orbit.

As with many satellite projects, slippages in the timetable began to occur. While for scientific satellites this type of problem did not always pose great problems provided the vehicle was ultimately launched, the delays on the TDF/TV-SAT project began to pose a threat to the whole programme rationale. This was because competition, in the form of a different type of television satellite being built for Luxembourg's Société Européenne des Satellites (SES) and capable of beaming 16 channels instead of just four, was threatening to make big inroads into TDF/ TV-SAT's potential market. The new satellite called Astra, an RCA 4000 orbiter built by GE Astro Space Division of the US, was less powerful than the Franco-German satellites but designed to be viewed using only slightly larger dishes, of the order of 60 centimetres in France, West Germany, southern England and the Benelux countries and 75 or 95 centimetres over much of Western Europe. SES was 33.3 per cent controlled by Luxembourg state banks, with a group of other banks holding most of the rest of the capital. Three British commercial television companies, London's Thames Television, the Plymouth-based TSW and Belfast-based Ulster Television, also took stakes in the venture.

DELAYS AND DIFFICULTIES

By mid-1986 it was becoming clear that the Franco-German project, far from having a comfortable period alone in orbit to prove its technology, would face competition from Astra at a very early stage, raising the possibility that French viewers, encouraged by their government to buy the small dishes capable of only receiving TDF-1 and TV-SAT, would find themselves deprived of the much wider choice of viewing available from Astra. A further complication had arisen because of the two Ariane launch failures in September 1985 and May 1986, which had led to a prolonged interruption in Ariane flights and thrown the launch schedules for all three satellites into doubt. Following earlier delays, TV-SAT 1 had been due to go into orbit in October 1986, TDF-1 in January or February 1987 and Astra in mid-1987. Now the launch dates were shrouded in uncertainty.

In early July 1986 Herbman Strub, head of the West German Research Ministry's space division, confidently announced the news of certification of flight-worthiness of TV-SAT 1 as 'a first goal, a first milestone ... in this joint French-German project'.[3]

But with no firm launch date, and a dispute raging between the regional governments of the West German states over who should broadcast on the satellite, its future looked far from bright.

In France the situation was if anything worse. The socialist government of Prime Minister Laurent Fabius had allocated franchises for the TDF-1 satellite which had been promptly revoked after the March 1986 elections by the incoming right-wing government of Jacques Chirac, following a dispute over the manner in which they had been distributed. Doubts also began to surface over the technical reliability of the TDF satellites, and in July the weekly magazine *Le Point* published an exclusive article quoting a note it said had been written by Posts and Telecommunications State Secretary Gérard Longuet to Chirac casting serious doubt on the wisdom of continuing with the project in its present form.

> The first of the second-generation satellites already ordered elsewhere in the world, which can carry up to 16 television channels, will be available in 1989, at the same time as TDF-2 ... the expected rental cost of a channel on this type of satellite is two to three times less than for TDF-1 and 2 ... the technical lead initially foreseen has today disappeared, and the technology of TDF-1 and 2 has even become a handicap

Longuet was quoted as writing.

> It therefore appears clear that both financial and technical considerations should lead to an immediate decision to abandon TDF-2 ... and if a decision is taken to launch TDF-1, it should be accompanied by necessary precautions to preserve this satellite's strictly experimental character and so avoid the large-scale equipping of homes with receiving dishes in conditions which could create severe problems for the future.[4]

On 29 July Chirac convened a meeting of senior government ministers at his elegant Hotel Matignon office on the left bank of the Seine. After a discussion examining the tough range of options a compromise was reached. TDF-1, which had been largely completed, would be launched as planned as an essentially experimental satellite, while not ruling out some form of commercial operations for it. Responsibility for TDF-2's future would be placed in the hands of a commercial company, under the auspices of Télédiffusion de France, to explore the possibility of funding the satellite privately.

Government spokesman Denis Baudouin, briefing journalists after the meeting, conceded that the perceived national interest in mastering this field of satellite technology was a factor in the satellite launch being given the go-ahead.

It is not a decision based purely on reason, it is a decision based on the determination of the government to be prcsent in an area where we have rivals ... the government has complete confidence in the technology which has been set in motion

he said. At the same time he made clear that ministers did not view the Franco-German high-powered satellite project as necessarily the only form of DBS system to be pursued in the future, noting that 'this project of direct broadcast by satellite is a complementary system to those of low and medium power'.[5]

MOUNTING COSTS

The meeting appeared to settle the project's future, at least temporarily. But in February 1987 the issue resurfaced when the right-wing French daily *Le Figaro* published excerpts of a letter written by Budget Minister Alain Juppé to Chirac, warning that continuing with the TDF satellite programme could 'be very damaging to state finances'. Juppé's gloomy prognosis was based on a government-commissioned report by the head of the state television channel Antenne-2, Claude Contamine, calling for a company to be set up to market the satellites, in which the satellite manufacturers, key potential operators and Télédiffusion de France would all take a stake.

The technical and financial aspects of the programme cannot be reconciled with its commercial exploitation ... as a result the financial risks of the project would be largely placed at the door of public finances

Le Figaro quoted Juppé as saying.[6]

The cost of the project had predictably risen from its original estimate, with the result that West Germany and France between them had spent close to $600 million on the joint satellite project, and the French Finance Ministry was becoming increasingly nervous over the prospect of further expenditure on a programme with an uncertain future.

Once more the French press weighed into the debate, pointing out that a project which had been designed as a technology leader was in danger of becoming obsolete, at least commercially if not technically, before it was launched. As columnist Albert Ducrocq wrote in the weekly French trade magazine *Air & Cosmos*:

It would be a great mistake to think that by abandoning the TDF-1 project French television is excluding itself from space. On the contrary it would do better to work with satellites which offer greater possibilities ... some of which have an advantage over TDF-1 in that they are already in orbit.

Both the Eutelsat and French Télécom telecommunications satellites already in space carried television channels, although not for direct broadcast to the home.[7]

There followed another meeting of senior ministers, again chaired by Chirac, on 25 February, which came out backing the TDF project but re-stating the government's determination that private industry should foot a large slice of future costs. The French state undertook to pay for the launch of TDF-1, guaranteeing up to Ffr 600 million to place it in orbit. And it accepted the conclusions of the Contamine report that a commercial company should be set up to fund the TDF-2 launch and the operating costs of both satellites. If private finance could not be found, TDF-1 would be launched by itself as a purely experimental satellite, since it would not be practicable for it to offer a commercial television service with no back-up satellite in orbit.

TV-SAT LAUNCH

Once more the dust settled, at least in public, and attention gradually switched back to the West German TV-SAT 1, whose launch date, after years of delays and political wrangling, was finally approaching. Set for Tuesday 17 November, an electrical fault with the Ariane-2 rocket forced a frustrating three-day postponement. The launch eventually took place on Friday 20 November, and was flawless – at least from the rocket's point of view. Within hours of going into orbit, however, it became clear that a serious problem had developed with the satellite. One of its two solar panels, on which it relied for electrical power, had failed to deploy properly. The panels, folded during launch, were due to unfurl in space to a total span of 19 metres. At first engineers were hopeful that the problem could be repaired. But as the weeks went by hope faded that it could be rescued, and by the end of February the satellite was written off. The failure was a major embarrassment, not only for the West German government which had invested so much of its money and reputation in the project, but for the whole Franco-German DBS programme. Financially it was a blow as well. Of the DM 390 million invested by the West German government in the satellite's construction and launch, only DM 95 million could be recouped through insurance.

TDF-1 LAUNCH

It was not until late October 1988, around four years behind schedule, that TDF-1 was finally launched. This time the solar panels opened, and by mid-November the satellite was in position and fully functional. Technically it was a success, but there was little rejoicing on the ground. Unlike most French space ventures, which take place against a background of official fanfare and media hype, TDF-1 had been branded – perhaps unfairly from the technical point of view – a white elephant. Prime Minister Michel Rocard described the whole programme as a scandal just weeks before the launch, criticising a decade of mistakes and the mis-spending of Ffr 2 billion of state funds. Even the subsequent successful launch of TV-SAT 2 on an Ariane 4 in August 1989 failed to completely dispel doubts about the wisdom of the programme.

Defenders of the TDF satellite project point out that ultimately its justification was to lay the groundwork for the development of high-definition television in Europe. The TDF-1 satellite broadcasts use the new D2-MAC Packet standard, which can transmit up to four million pixels, or points of light, onto a screen, more than ten times the 300 000 pixels possible under the traditional PAL system used throughout most of Western Europe. D2-MAC is therefore a step in the development of high-definition television, ushering in the era of advanced flat television screens expected to replace the more traditional bulky cathode ray television sets by early in the next century.

In contrast to TDF-1, which was ultimately launched primarily as a technological spacecraft although broadcasting a selection of programmes including a Franco-German cultural channel, the concerns of the operators renting channels on Astra were more directly concerned with profits and the demands of the market. Astra followed TDF-1 into space aboard the very next Ariane flight, the first commercial launch of Ariane-4 on 10 December 1988, and was successfully placed in its correct orbit and declared operational within a few weeks.

ASTRA

The fact that Astra had made it into orbit was in itself a political victory for SES, the Luxembourg company which owned it. For the entire Astra direct broadcast project, which represented a commercial challenge not only to TDF but also the Eutelsat satellites, had been strongly resisted by the Eutelsat organisation, partly on technical grounds. Eutelsat director-general, Andrea Caruso, in an interview in Paris the day before the Astra launch, forcefully explained his organisation's point of view.

> Astra ... sell their satellite as a broadcasting satellite. It's not. It is a satellite which operates in frequency bands which are not for broadcasting, which are for the distribution of television. Broadcasting is for

reception by everybody, with a small antenna on the roof. Distribution is to distribute to antennae which then rebroadcast via the land lines, via the microwave system...[8]

Caruso also attacked the involvement of British Telecom, a Eutelsat signatory, in the Astra project through its March 1987 agreement to set up a joint marketing venture with SES to market 11 of Astra's transponders to British-based programmers. 'BT is acquiring a de facto monopoly in those satellite systems where they did not have a de jure monopoly ... BT retains now a de jure monopoly in Intelsat and Eutelsat and Inmarsat and a de facto monopoly in Astra', Caruso said.[9]

British Telecom officials rejected the accusation, insisting that its association with Astra was motivated by a lack of available space on the Eutelsat system.

This isn't a regulated monopoly, we've had strong competition.... It was simply a question of timing. It was only because Eutelsat was unable to meet our customer requirements that pushed us down the route to securing Astra capacity

Steve Maine, Head of BT's broadcast and visual services, commented, adding

I don't see our involvement in Astra as incompatible in any way with our investment interest in Eutelsat ... we attached supremacy to meeting the requirements of UK-based customers. We attach more importance to that than to maximising our investment in the Eutelsat system.[10]

Despite the polemic involving Eutelsat, SES succeeded in getting its satellite airborne, and within two months came the splash inauguration in Britain of Rupert Murdoch's Sky channels, all beamed from the satellite.

SKY TELEVISION

These comprised Sky Channel itself, a revamped version of an entertainment channel of the same name which had been available to cable subscribers via the OTS-2 satellite since 1982, a 24-hour news channel called Sky News, a film channel, Sky Movies, and an 18-hour-a-day sports channel called Eurosport, as well as a Disney channel. As part of the drive to win viewers, Murdoch had forged a deal with Alan Sugar of Amstrad Consumer Electronics to manufacture and market receiving dishes at £199 in Britain, plus a £40 to £70 installation cost, bringing the technology within reach of a large section of the population. Posters across Britain in the weeks preceding the start-up proclaimed 'You'll never say there's nothing on the box again', and two further channels were promised before the end of 1989, one comprising Disney entertainment material and the other Sky Arts, billed as a 'television treasure house of arts with hours and hours of sumptuous quality entertainment'.[11]

The investment was huge, with expenditure of up to £150 million expected in the first year of operation, in addition to the £40 million lost on Sky Channel during its earlier six-year life and £25 million invested to upgrade it for the Astra launch. 'This is the television revolution', Murdoch proclaimed to his viewers on the opening night – an audience estimated by the British media at little more than 50 000 homes, despite Sky's claim to be available to more than 600 000 homes in Britain and Ireland through cable or dishes.

Murdoch could claim the satisfaction of owning the first operational direct-broadcast satellite service in Europe. However, critical reaction to the Sky start-up was mixed, with some British newspapers welcoming the new service but others focussing on the high proportion of American series and film repeats on offer. 'Sky can only be described as expanding choice if you believe that the historical failure of British television has been to provide enough sport, old movies and soap operas', television critic Mark Lawson wrote in the London *Independent*. Entitling his piece 'The moronic inferno', he also complained about the quality of the news channel:

> Sky News Tonight and its morning variant Sky News on the Hour employ some decent middle-ranking BBC hands . . . but so far betray the sense that the running orders are dictated not by the day's events but by the sentimental agenda of the tabloid press.[12]

Viewers, according to the Sky adverts, could choose between such items as the NBC Nightly News, taken on a live feed from New York in the early hours of the European morning, and 'a lather of soaps' on the entertainment channel.

Apart from Sky, other viewing options on Astra included two entertainment channels run by British book chain W. H. Smith.

Murdoch's chief aim in making such a splash with the start-up of the family of Sky channels was to gain a firm foothold in the British market, as well as in continental Europe, before the rival British Satellite Broadcasting (BSB) system could start transmitting its three satellite channels in September 1989. BSB's target date for full commercial start-up was subsequently delayed until early 1990, giving Murdoch a further competitive edge.

BSB was a company owned in part by the Pearson publishing and media group, the Bond Corporation and Granada Television, aiming ultimately to broadcast a total of five channels including a diet of films and news. Unlike Murdoch, who had decided to broadcast on the existing PAL system to reach the maximum audience as quickly as possible, BSB had opted for the more advanced D-MAC technology. Following the joint

Murdoch-Sugar announcement of a £199 satellite dish in July 1988, BSB unveiled a lightweight flat antenna the following month which it said would retail in the UK for about £250, stiffening the competition for viewers while also adding to consumer confusion over the merits of the different available systems.

> We believe the consumer will demand two things in choosing a satellite service. The first is good quality programmes focussed on British tastes and the second is technology which will adapt to television innovations rather than being chained to the technology of the Sixties

BSB chief executive Anthony Simonds-Gooding declared.[13]

COMMERCIAL COMPETITION AND TECHNOLOGICAL CONFUSION

The battle between BSB and Sky for viewers assumed wider significance because of the technological battle concerning broadcasting standards which lay beneath the rivalry. But to the consumer, whose main concern lay simply in receiving a good-quality picture rather than with its means of delivery, the argument seemed esoteric, reminiscent of the similar incompatibility of video recording systems when they had been first popularised on the market ten years earlier.

Aside from the confusion over technical standards and equipment, made worse by the fact that Japanese manufacturers were also working on experimental high-definition television equipment which would be incompatible with that being developed in Europe, there was also the inevitable uncertainty associated with a fledgling television service that relied on a certain audience penetration to succeed, and yet had to start operations with a considerably smaller segment of the market. Sky forecast that it would be reaching upwards of 1.5 million homes in its first year of operations, while BSB predicted it would be reaching three million homes in Britain within two years and as many as 18 million over the longer-term, equivalent to 90 per cent coverage of the population. Independent consultants were more cautious, predicting that around two million homes would be able to receive satellite television in Britain by late 1991.

By late February, the same month as Sky took to the air and less than three months after Astra was launched, SES felt sufficiently confident of its commercial prospects to arrange the purchase of a second satellite, potentially doubling its capacity to 32 channels. Because it succeeded in finding a GE Astro satellite that had already been built for another purpose and then moth-balled, it was able to arrange for a launch by late 1990, much sooner than the three-to-four year wait that would have been necessary had it ordered a satellite to be specially constructed.

The European skies appeared likely to become rapidly crowded with direct-broadcast satellites. The French satellite TDF-1, in orbit since late 1988, was due to be joined in space by its West German counterpart TV-SAT 2 in the second half of 1989 and by the second French satellite TDF-2, assuming its launch could be funded, in early 1990.

Apart from the TDF/TV-SAT and Astra programmes, another DBS satellite took to the sky in early 1989. Ariane flight V30, using the last of the Ariane-2 rockets, blasted the Swedish Space Corporation's 2.1 tonne Tele-X satellite into orbit on 1 April. The vehicle was designed to operate as a hybrid DBS and business communications satellite, with two direct-broadcast television transponders and two transponders allocated for business communication channels. Two back-up transponders were also aboard, one each for DBS and business communication use in case of technical problems. Tele-X was a derivative of the TDF/TV-SAT programme, built by Eurosatellite in Munich, and the entire Tele-X system was owned by Nordiska Satellitaktiebolaget (the Nordic Satellite Company), in which the Swedish government had an 85 per cent stake and the Norwegian government 15 per cent. Under a sub-agreement the Finnish government took over a 3 per cent stake from Sweden.

CONCERNS OVER REGULATION

Amid the phenomenal expansion of television capacity, which would bring more than 40 new channels within reach of European viewers used to having access to only four or five, governments began to consider ways in which the industry could best be regulated and whether some of the satellite capacity could be used for state-promoted channels to improve cultural links. In May 1988 a committee on the future of television, headed by former French President Valéry Giscard d'Estaing, called for the establishment of a European satellite television service, funded both by governments and television organisations, to produce programmes and news of high quality. The annual cost of such a service was estimated at £70 million. The committee also called for programme quotas to ensure a continued outlet for European-made programmes on satellite channels in the face of competition from often-cheaper US material. 'In all there are 32 television services available via satellite in Europe, but we consider there is no truly European service broadcasting in the public interest', Georg-Michael Luyken, the committee's executive secretary, said.[14]

The committee also proposed tax incentives for people to invest in making programmes and suggested that a new organisation called the European Television Forum should be set up to fulfil the role of a self-regulatory body for both private and public broadcasters, in the absence of enthusiasm from governments for a more rigid, pan-European authority imposing regulations on broadcasting.

In March 1989, following the TDF-1 and Astra launches and the failed launch of TV-SAT 1, Trade Ministers of the 12-nation European Community met in Brussels to approve a draft package of proposed regulations on television hammered out by officials in smoky committee rooms over the preceding months. This laid down guidelines for advertising, set an indicative quota for European content in programming and defined limits for sex and violence on screen. The regulations were to apply to all forms of television broadcasts within the Community, not just satellite transmissions, in order to prevent individual member countries from using discrepancies between national and community law as an excuse for blocking satellite transmissions on their territory.

France had waged a campaign to have a mandatory minimum quota of 50 per cent European content placed on programmers, but this met opposition from many other countries, partly on philosophical and partly on practical grounds, and so became diluted in the negotiating process to an instruction to television stations simply to set aside a majority of their transmission time if feasible to European programmes.

Another regulatory issue was raised by an agreement that Sky Television's Eurosport channel had reached with several leading European broadcasters, providing the channel with exclusive access to top sporting events. The European Commission formally objected to the arrangement, saying that it posed a threat to competition within the Community. The dispute was an indication of the growing complexity of regulating television across boundaries which had, for all practical purposes, ceased to exist.

As the 1980s drew to a close, private satellite television had established itself as a small but growing part of the European television scene, forcing the authorities to rethink the old regulations designed for a vanishing era of state-regulated, national channels which stopped at frontiers. Public interest, or at least curiosity, had been aroused in Britain by the brash marketing campaign leading up to the Sky launch, while in France and Germany the national debate was focussing on the form that satellite broadcasting should take, rather than whether it was desirable in principle. The one thing both critics and enthusiasts could agree upon was that satellite broadcasting had arrived, and could not be ignored.

12 The French Manned Space Programme

Spring was turning to summer on the Russian steppe, wild flowers punctuating the grey-green monotony of the wide horizon, as Soyuz T-6 roared into life on the pad at Baikonour cosmodrome, Kazakhstan and thundered skyward. The date was 24 June 1982. For the Soviet Union it was a relatively routine flight, just another in a long series of missions to orbiting Salyut space stations which had been gathering pace over the previous ten years. But for France it signalled its entry into the élite club of nations which had sent their citizens into space. Jean-Loup Chrétien, a 43-year-old test pilot, was on board alongside two Soviet cosmonauts, mission commander Lt.-Col. Vladimir Alexandrovich Dzhanibekov, a veteran of two previous space flights, and flight engineer Alexander Sergeyevich Ivanchenkov. Chrétien's role was officially to perform scientific experiments. But it was also to notch up a prestigious landmark for the French space programme – becoming the first West European astronaut in space, and in the process beating into orbit West German Ulf Merbold, who was at that time undergoing training in the US for a Spacelab flight aboard the shuttle under the auspices of the European Space Agency.

ORIGINS OF FRANCO-SOVIET SPACE CO-OPERATION

Chrétien's historic flight provided dramatic evidence of France's determination to pursue a vigorous space programme of its own as well as playing a dominant role in the European Space Agency. In fact Franco-Soviet space co-operation had proved particularly fruitful and long-running, going back 16 years to an accord concluded by President Charles De Gaulle during a visit to Moscow in June 1966 which pledged bilateral co-operation on the study and exploitation of space for peaceful purposes. The agreement covered such areas as the exchange of scientific information, development of meteorology from space and early experimental work on colour television transmissions by satellite.

In 1967 joint Franco-Soviet work started on the study of the polar thermosphere, involving firing MR-12 meteorological rockets from Heyss Island in the Arctic territory of Franz-Josef-Land to measure the atmospheric temperature by releasing artificial clouds of sodium. The following year saw the start of a series of OMEGA experiments using French high-altitude scientific balloons. These were released from the French archipelago of Kerguelen, in the southern Indian Ocean, and the

Archangel area of the Soviet Union, and took a variety of measurements relating to X-rays, the ionosphere and the Earth's magnetic field. Later in 1969 studies started on the chemical composition of the upper atmosphere, using French Dragon research rockets carrying Soviet radio spectrometers launched from a base in the Landes region of southwestern France.

The year 1969 also saw the delivery to the Soviet Union of two French-made laser reflectors, commissioned under a 1967 accord, to be mounted on the Soviet unmanned lunar exploration vehicle Lunokhod-1. This was an eight-wheeled robot explorer, resembling a large iron bathtub on wheels, designed to be landed automatically on the moon and then to move slowly around taking scientific measurements. Lunokhod-1 was placed on the moon by Lunik 17 in 1970, the year after the first American manned landings, and travelled 10 kilometres in nearly a year, analysing the lunar soil and magnetic field. Its sister craft Lunokhod-2 joined it in January 1973.

The role of the French laser reflectors was to measure the precise distance from the Earth to the moon, using laser readings of the Lunokhod's position from telescopes in the Soviet Crimea, in Pic-du-Midi, France and in the United States. Laser beams were bounced over the half-million miles to the Lunokhod and back, enabling scientists to determine to within less than a metre the distance from the Earth to the Moon. The Lunokhod experiment was also used to shed light on such phenomena as continental drift, tidal movements and the rotation of the Earth. The travels of the twin Lunokhods across the moon's surface produced large quantities of scientific data, but were inevitably overshadowed in the Western media by the American Apollo moon landings during the same period.

In 1971 French researchers had a solar radio emission experiment aboard the Soviet interplanetary probe Mars-3, launched in late May of that year. In July 1973 the Soviet Mars-5 probe was launched, again with French instruments on board to study solar radiation, while Mars-6 and Mars-7 carried similar French experiments. In addition, Franco-Soviet experiments on board Mars-5 and Mars-7 studied ultra-violet radiation in the upper atmosphere of Mars.

In December 1971 a Franco-Soviet satellite, Auréol, had been launched to study the upper polar atmosphere and also the way in which aurorae borealis are formed, followed by similar satellites over the next ten years.

There was a series of joint meteorological experiments involving Soviet sounding rockets in 1973, followed in 1974 by the release of sounding balloons carrying French and Soviet instruments from a base at Kiruna, Sweden. In 1975 French Eridan rockets were launched to study ionospheric disturbances and in June of that year a Soviet rocket launched a French scientific satellite, SRET-2. Later in 1975 the two countries conducted biological experiments aboard the Soviet satellite Cosmos-782, studying the effects of cosmic radiation on primitive life forms.

The year 1977 saw the launch of another French satellite, Signe-3, by the Soviet Union, this time to study X- and gamma-radiation and ultra-violet solar radiation. In August the Soviet satellite Cosmos-936 carried another series of Franco-Soviet biological experiments into orbit, aimed at studying ways of helping astronauts avoid the harmful effects of cosmic radiation during long-duration space flights. And the following month the Soviet satellite Prognoz-6 was launched to study solar radiation, again with French instruments.

In 1978 French biological experiments were included for the first time in the work being done aboard the Soviet space station Salyut 6, a reflection of France's growing interest in developing a manned space programme of its own. The joint Franco-Soviet Cytos experiment was aimed at studying simple cell growth in weightless conditions, compared to their performance on Earth. The experiment was conducted on a flight which, for a different reason, was making its own piece of space history. For when the two-man Soyuz-27 crew took off on 10 January 1978 to dock with the Salyut space station, they were acting as a resupply mission for another two-man crew who had been launched on Soyuz-26 a month earlier and were to stay in orbit for three months. It was the first time that two spacecraft had been docked simultaneously to an orbiting space station, and marked a new step forward for the Soviet manned programme. The commander of the Soyuz-27 mission, Dzhanibekov, was later to take charge of Jean-Loup Chrétien's flight.

The Cytos experiment was successfully repeated the following year by cosmonauts Vladimir Afanasevich Lyakhov and Valeri Viktorovich Ryumin on their record-breaking Soyuz 32 flight to Salyut 6. The two cosmonauts spent 175 days in space, and during their stay took delivery of some biological experiment packages prepared by laboratories in Bordeaux, Meudon and Grenoble, which had been launched from Baikonour aboard the unmanned supply craft Progress-5.

French instruments were once again aboard the Soviet interplanetary probes Venera-11 and Venera-12, launched within five days of each other in September 1978, which continued earlier investigations into gamma radiation and took measurements of Venus with an ultra-violet spectrometer during a fly-past to analyse its upper atmosphere. In September 1979 another Soviet satellite with French biological experiments aboard, Cosmos-1129, went into orbit, while in mid-1980 there were more launches of Soviet meteorological rockets from the French base on Kerguelen.

1981 saw a series of tests concerning television transmission by satellite between France and the Soviet Union, while towards the end of the year two more probes, Venera 13 and 14, were launched with French, Soviet and Austrian instruments to measure X- and gamma radiation and magnetic fields.

Therefore when the Soviet Union proposed to France that it should nominate a candidate to fly aboard a Soviet space mission, it was not an empty political gesture but the fruit of well over a decade of co-operation in space science.

EARLY PREPARATIONS FOR THE FIRST FRANCO-SOVIET FLIGHT

The offer was made in April 1979 and immediately accepted by France, and by early 1980 a total of 430 candidates had applied to the French space agency (CNES) for the seat. A series of rigorous background and medical tests whittled this number down to five candidates by the beginning of March, four men and a woman. More tests followed before the final choice was made of two candidates to undergo nearly two years of training, Jean-Loup Chrétien and Patrick Baudry, both of whom were test pilots.

When they arrived at Star City, the Soviet cosmonaut training complex near Moscow, on 8 September 1980, it had not yet been decided which of them would go into space and which would be designated the reserve. While the Soviet Union required two Frenchmen to undergo training in order to have a backup should one fall ill or be forced to withdraw for some other reason, it was also of great value to the French space authorities to have two astronauts in training. After the flight the two would form the backbone of a future core of French astronauts, destined ultimately to fly French or European manned spacecraft into orbit.

Chrétien, the older of the two, was born on 20 August 1938 in La Rochelle on the French Atlantic coast and had been a fighter pilot all his professional life. He graduated from the French Ecole de l'Air (flying school) in 1961 and flew Super-Mystère B2 and Mirage-3 fighters before joining the test pilots' school at Istres in 1970 and later becoming chief test pilot on the French Air Force's Mirage F1 programme. An Air Force colonel, he lived in Brittany with his four sons. His greatest passion in life, apart from flying, was playing the organ, but he was also a keen yachtsman and golfer.

Baudry, eight years his junior, was born in Douala, Cameroun, and spent many of his childhood years in Africa. Smaller than Chrétien, dapper and slimly built, Baudry graduated from the Ecole de l'Air in 1970 and completed his training as a test pilot in England eight years later, subsequently flying Jaguar aircraft and rising to the rank of major in the French air force. Married with a young daughter, he was based in Bordeaux.

In a search to find a French title which would distinguish them from

American astronauts and Soviet cosmonauts, Baudry and Chrétien were officially described as spationautes. France was determined to demonstrate its independence in space policy, even at the apparently mundane level of vocabulary.

The training programme on which they embarked was arduous in the extreme, and focussed on three different aspects of the flight. Firstly they had to learn how to fly the Soyuz spacecraft and the new Salyut-7 space station which one of them would be visiting. Secondly they had to master the scientific experiments that would be conducted during the week-long flight. And thirdly they had to prepare themselves physically for the duress of the flight.

SALYUT-6 AND -7 SPACE STATIONS

By the time that Chrétien and Baudry arrived at Star City, the Soviet authorities had become accustomed to international trainee cosmonauts. For the Soviet Union had organised an intensive three-year programme of Intercosmos visits to the Salyut-6 space station by cosmonauts from no less than nine communist countries, starting perhaps surprisingly – in view of Moscow's strained relations with Prague over the years – with Czech cosmonaut Vladimir Remek on Soyuz 28 in March 1978. Over the following months cosmonauts from Poland, East Germany, Bulgaria, Hungary, Vietnam, Cuba, Mongolia and Romania all flew to the Salyut space station. In addition to the Intercosmos programme, the Soviet Union had also had to deal directly with American astronauts during preparation for the July 1975 docking of an Apollo and Soyuz spacecraft, which involved two days of largely ceremonial handshakes and televised press conferences from space.

It was decided from the beginning that both Chrétien and Baudry would have to speak Russian, since organising interpreters for them for the two-year training period and the intensive flight would have been very time-consuming and inefficient. Therefore, before leaving Toulouse for Moscow, both men were subjected to two months of total immersion classes, confined to language laboratory booths for up to 12 hours a day.

In the Soviet Union their training simulated all aspects of the space flight, covering everything from life on board Salyut to how to cope with an emergency splashdown in the sea if the capsule missed its target descent area in Soviet central Asia. They undertook a series of training flights in simulated weightlessness aboard a specially-prepared IL-76 aircraft with a padded fuselage, and practised camping out on the frozen tundra to prepare for the eventuality of having to spend days out in the bitter cold if rescue teams had difficulty reaching them after touch-down.

Chrétien and Baudry arrived in Moscow at a busy time in the space programme, both in the Soviet Union and elsewhere. Back in Western Europe the Ariane rocket had been temporarily grounded after the triumph of its first flight had turned sour with the failure of its second, and work was under way to put the rocket back on the launchpad. Meanwhile the Soviet manned programme was continuing at a fast pace. In the six months following their arrival at Star City there were no fewer than four Soyuz flights to the Salyut-6 space station, including one lasting two-and-a-half months. But while the flights received routine coverage in the Soviet media, especially those involving politically-significant guest appearances by foreign cosmonauts, they barely rated a paragraph in the Western press. What did capture the Western public's attention, completely upstaging the Soviet programme, was the spectacular first flight of the US shuttle in April 1981. For three days the newspapers spoke of nothing else, and the United States, after having laboured in the shadow of the Soviet space programme since the end of the Apollo moon flights more than eight years earlier, appeared to be forging ahead again in the space race.

There followed a final Soyuz flight to Salyut 6 and two more space shuttle flights before the new Salyut-7 space station was blasted into orbit on 19 April 1982. By this time Jean-Loup Chrétien had been selected as the French cosmonaut to go into orbit, with Baudry to remain on the ground. The launch of the new Salyut station marked the start of the final preparatory phase for Chrétien's flight, since this was the vehicle on which he was due to spend most of his week in orbit.

The new station was different from its predecessors in several respects, notably that it was specifically designed for long-duration flights with a wide variety of scientific equipment on board, more modern, automatic flight instruments and a range of docking facilities permitting visiting manned craft and unmanned resupply vehicles to dock simultaneously. While still far from luxurious, it was brightly painted and intended to be more comfortable for the cosmonauts. Space suits had also been re-designed to take account of the changed environment.

The space station was 15 metres long, 4.15 metres in diameter and nearly 17 metres across with its solar panels fully extended, with a mass of around 20 tonnes.

It was divided into five compartments, separating the machinery necessary for maintaining the vital systems of the craft from living and working areas. It still had relatively rudimentary shower and toilet facilities but also featured hot and cold running water, a fridge and an exercise area to help cosmonauts fight boredom and muscle deterioration.

On 13 May Soyuz T-5 lifted off from Baikonour. This was one of the new-generation Soyuz craft carrying two cosmonauts, Anatoli Berezovoi and Valentin Vitalyevich Lebedev, who were to become the first long-stay visitors aboard Salyut-7, setting a new space endurance record by staying in orbit until December. The following month, after 21 months of hard training, it was finally Chrétien's turn to go into space.

CHRETIEN'S FLIGHT

On the evening of 24 June the weather was fine at the Baikonour cosmodrome as the three-man crew of Soyuz T-6 ran through their pre-launch checks with ground control. Dzhanibekov, Ivanchenkov and Chrétien were seated in the capsule atop the A2 rocket, while the back-up crew of Leonid Kizim, Vladimir Solovyov and Patrick Baudry were at ground control, monitoring developments and standing by to make suggestions or offer advice if problems arose.

Lift-off was flawless and the nine-minute flight into orbit passed off without a hitch. Some preliminary work, including ensuring that the capsule was correctly oriented, was conducted in the first few orbits, followed by a ten-hour sleep period prior to the docking manoeuvres with the Salyut space station. After completion of the manual docking on the night of June 25, the three-man Franco-Soviet crew crawled from their Soyuz craft through the hatch into the Salyut space station to a warm Russian welcome of bear hugs from the resident crew. Official congratulations then began arriving, Soviet leader Leonid Brezhnev transmitting his congratulations live to the orbiting crews while French President François Mitterrand contented himself with a telegram.

During the following seven days the Franco-Soviet crew propelled themselves through a crowded programme of scientific experiments which fell into three broad categories, namely material science, astrophysics and bio-medicine. The material science experiments included work on a newly-designed oven on Salyut which could reproduce temperatures up to 1000 degrees Celsius, and investigations of the structure of liquids. Bio-medical experiments included studies of the interaction between human senses and the body's motor system, an analysis of the impact of antibiotics on bacteria in space and further work on the effect of heavy ions, part of the studies conducted by the two countries over a number of years on the problem of radiation in space flights.

The bio-medical experiments also included a French-developed echograph to study the cardiovascular system of cosmonauts in weightlessness, including the way blood circulates through the body under micro-gravity. Chrétien used himself as guinea-pig for this experiment, and so successful was it that scientists requested that the space station's two permanent crew should continue the experiment upon themselves in the months following Chrétien's departure.

The astrophysical experiments included photographic work on faint sources of visible and infra-red light in the Earth's atmosphere and interplanetary space, and studies of cosmic X-rays.

While Chrétien's experiments provided useful data, there was some frustration among scientists that it was not possible to start the experiments earlier in the flight, due to his dual role as scientific experimenter and co-pilot of the Soyuz spacecraft. Since the experiments were designed to monitor the adaptation of the human body to weightlessness, the long period between launch and Chrétien actually entering the space station represented a gap in data.

One of the principal findings from Chrétien's experiments was that the rate of blood flow around the heart accelerated in microgravity, while remaining unchanged in the more extreme upper and lower areas of the body.

After 125 orbits docked to Salyut-7 Chrétien, Dzhanibekov and Ivan-chenkov separated their Soyuz craft and put themselves on a re-entry trajectory, landing in the early evening of 2 July. The mission was promptly declared a spectacular success by both French and Soviet officials involved, above all on the technical level where the entire programme of experiments appeared to have been completed without a hitch.

The following few days brought the obligatory official receptions, including medal ceremonies at both the Kremlin and the Elysée palace. Evaluation of scientific data took many months, and in many cases produced ideas for follow-up experiments on subsequent flights.

SHUTTLE INVITATION

While France had won the race to have one of its citizens become the first West European astronaut, within three years it was to pass another notable landmark, flying its astronauts on both the Soviet and US space programmes. Chrétien and Baudry were soon packing their bags for Houston, Texas, having been accepted for training as payload specialists on the US shuttle programme. This time Baudry was to be prime astronaut and Chrétien back-up. NASA had invited the French aboard the programme because its interest had been aroused by the time and resources France was putting into research on long-duration space flights, and particularly the work done by Chrétien aboard the Salyut space station.

The flight to which Baudry was assigned, shuttle mission 51-G aboard Discovery, was to prove one of the busiest shuttle missions to date. The agenda included the launch of three telecommunications satellites and a range of scientific experiments as well as the first attempt to test President Reagan's controversial Strategic Defense Initiative (SDI) technology in space. Apart from the five Americans aboard, there would be Baudry, whose prime function was to conduct medical experiments, and a Saudi prince, Sultan Abdul Aziz Al-Saud, who had been allocated a seat because his country had paid for the launch of one of the communications satellites, Arabsat.

BAUDRY'S FLIGHT

Lift-off from Cape Canaveral came at 07.33 local time on 17 June 1985, almost three years to the day after Chrétien's flight from Baikonour. Violent thunderstorms had battered the Cape the previous afternoon, putting the flight schedule in jeopardy, and just four hours before launch a power cut hit the pad. But Discovery's main engines ignited right at the start of its launch window, smoke billowed from the solid-fuel boosters and the shuttle climbed away from pad 39A into its pre-ordained orbit.

The commercial part of the flight was the deployment of the three communications satellites. These were Morelos A, a Mexican satellite built by Hughes Aircraft Co, Telstar 3 owned by American Telephone and Telegraph Co (AT&T) and Arabsat. Arabsat had been commissioned by the Arab Satellite Communications Organisation and was designed to improve telephone and data transmission links across the Arab world from North Africa to the Gulf, over long distances where land lines were either too expensive or impractical and where satellite communications offered the chance to quickly upgrade the standard of service. There was French involvement in the Arabsat project, since the vehicle had been jointly developed by Aérospatiale with Ford Aerospace of the US.

All three satellites were released from the shuttle's cargo bay and subsequently reached their correct orbits, by no means a foregone conclusion in view of several satellite losses that occurred during the mid-1980s.

Apart from Baudry's series of medical experiments, one of the main goals of the flight was the first ever test in orbit of SDI technology, universally referred to in the press as Star Wars. President Reagan's vision of a space-based defence system rendering nuclear weapons obsolete by early in the next century had provoked a massive debate among governments and in the Western press, arousing strong opposition due both to doubts over whether such a system could ever be reliable and due to its huge costs, estimated at $25 billion or more.

The Strategic Defense Initiative Organization (SDIO) had been set up to oversee the experimental development of the system, and had booked shuttle mission 51-G to conduct its first test of a laser tracking system in space.

The test called for a low-powered laser beam to be aimed from Maui island, Hawaii, skywards to Discovery orbiting overhead, bounced off an

8.5-inch diameter retroreflector attached to a window on the left side of the shuttle and picked up again in Hawaii.

While the concept was relatively simple, putting it into practice proved to be more of a challenge. The first attempt was made on 19 June, two days into the flight, but an error in feeding data into the shuttle's computer resulted in Discovery facing completely the wrong way as it headed across the Pacific towards the test site, its reflector aimed out to space rather than facing the ocean below. Tempers rose both on the flight deck and at mission control as the crew and ground engineers blamed each other for the mess. A NASA post-mortem subsequently concluded that the crew had not been responsible for the errors.

Although the full experiment had to be postponed, ground engineers decided to go ahead with testing the accuracy of the laser by pointing the beam at the shuttle as it flew over, despite the craft's temporary inability to reflect it back again. The beam, which was a quarter of an inch across when it left Hawaii, had expanded to 30 feet across when it reached the orbiter, bathing it in a blue-green light which the astronauts inside could clearly see.

Two days later, on 21 June, the computer data was correctly fed in, the shuttle came over the Pacific with its reflector pointing downwards and the beam found its target. It tracked Discovery for two and a half minutes until it flew out of range, attaining a high degree of accuracy despite strong winds on Hawaii and the resultant atmospheric turbulence which had threatened to disrupt the experiment.

While work on the laser experiment was continuing, Baudry was putting himself and fellow crew members through a demanding series of medical tests to fulfil his primary role on the mission. The experiments were a direct follow-on from the tests performed by Chrétien on Salyut-7 three years earlier. As guinea pigs he used himself, the Saudi payload specialist Al-Saud and the 42-year-old US mission specialist Shannon Lucid. The three had been thoroughly tested a month before the launch, and again just a few days before lift-off, in order to give scientists reliable data with which to compare the results achieved in weightless conditions and after their return to Earth.

One of the main aims of Baudry's experiments, primarily using an echocardiograph, was to study the flow rate of blood in the central region of the body, to supplement similar data gathered by Chrétien on other parts of the body. Baudry was able to start his work within three hours of launch, compared with the delay of nearly 48 hours in Chrétien's case because of the long transitional flight phase in the Soyuz craft before reaching Salyut. As a result Baudry's data tracked changes in the body from a much earlier stage after reaching weightlessness. Related experiments included investigations of the causes of vertigo and space sickness, the role of the senses in helping the body to orientate itself to microgravity conditions, and ways in which equilibrium is maintained.

The equipment used by Baudry on the shuttle was little different from that used by Chrétien on his flight, although some modifications were necessary. Chrétien had left instruments aboard Salyut-7 to enable crews to continue with the experiments, both to provide scientific data and also as a means of monitoring their own physical condition.

At the same time as Baudry was orbiting the Earth in Discovery, his old friend Vladimir Dzhanibekov, mission commander on the Franco-Soviet flight, was back aboard Salyut-7, although in very different circumstances to those in which Chrétien had visited. Early in 1985 the station had suffered technical problems which put it out of action, while leaving it superficially undamaged and still in orbit. Dzhanibekov's unenviable task, with flight engineer Viktor Savinykh, was to dock a Soyuz craft to the station and reactivate it, living inside it for nearly a week while they mended its life support systems. So it was that coincidentally in June 1985 France had two sets of bio-medical instruments in orbit, one aboard Discovery with Baudry and one being resuscitated aboard Salyut-7.

In addition to Baudry's medical experiments, the 51-G shuttle mission was also responsible for testing a free-flying platform loaded with scientific experiments called Spartan 1. This was a small platform, stowed in the shuttle's cargo bay, which was pushed out into an independent orbit for around 45 hours, during which it flew nearly 200 kilometres apart from the orbiter. Its task was to carry out astronomical observations on the core of the Milky Way and on a group of galaxies known as Perseus, recording data on tape for subsequent analysis on the ground.

Apart from the scientific value of any data gathered, the intention was also to test the working of the platform, which it was hoped could provide a model for other platforms to ultimately replace sounding rockets for astronomical observations, vastly increasing the time over which scientists could gather data. Spartan's performance was hailed enthusiastically by mission controllers as living up to its pre-flight expectations.

The shuttle mission lasted just over seven days, and at 06.12 local time on 24 June touched down on the lakebed at Edwards Air Force Base, California, an hour after commander Charles Brandenstein had fired the engines to swing it out of its orbit. It had been originally due to land at the Kennedy Space Centre, Florida, but the landing site was switched due to concern over the shuttle's braking system. The early morning landing went smoothly, apart from a minor problem as the vehicle came to a halt, when the main left wheels sank several inches into the lakebed. No damage was sustained, and it was towed clear without difficulty.

FRENCH COMMITMENT TO MANNED SPACE FLIGHT

With two astronauts successfully sent into orbit, and valuable experience gained of both the Soviet and US manned space programmes, the French space authorities were anxious to keep the momentum up. The French space agency CNES, unlike some of its European counterparts, enjoyed firm support across the political spectrum and could count on generous government grants to fund ambitious programmes. Nevertheless, especially in the wake of the Challenger accident which occurred just six months after Baudry's flight, there were some voices of mild concern raised in France over the wisdom of placing too much emphasis on putting French astronauts into orbit.

The French Academy of Sciences, in a report on space policy published in March 1988, said there were some tasks done by astronauts which could be performed by robots, and while arguing that man did have a role in certain circumstances in space, warned that it would be a mistake to divert precious resources from space science programmes into achieving 'space spectaculars' involving manned missions.

> The financial effort which France will be putting into the manned space programme should not in any way jeopardise the budgets of other research sectors. It would be irresponsible to sacrifice scientific research, and particularly basic research, in order to conduct an ambitious and spectacular space policy

it said.[1]

Yet despite such gentle attempts to start a debate over the wisdom of spending millions of dollars sending men into space, the French political class and media remained virtually unanimous in their endorsement of the need to have French astronauts in orbit.

> Putting man in space is a key ambition for France and for Europe ... this ambition is in tune with the aspirations and capabilities of European nations

Prime Minister Jacques Chirac told assembled aerospace dignitaries in June 1987.[2]

That attitude, shared by both Chirac's right-wing supporters and the French Socialists, remained a cornerstone of the country's space policy through the 1980s.

CHRETIEN RETURNS TO STAR CITY

By late 1986 France and the Soviet Union had agreed to prepare for a second joint flight, in which a French astronaut would fly to the new Mir space station. The accord was a clear affirmation of France's determination to gain further experience in manned space flight. This time the mission would last a month, four times as long as Chrétien's first flight. Two astronauts were selected for the training process, Chrétien once more and Michel Tognini, a 37-year-old test pilot in the French air force. Chrétien would get the flight, while Tognini would train as back-up.

The training programme that they followed at Star City had similarities to the one that Chrétien had undergone with Baudry, but also some key differences. This time the destination was the space station Mir, launched in February 1986 and designed to become the cornerstone of Soviet space science research until well into the 1990s. The core module was just over 13 metres long and up to 4.2 metres in diameter, with a mass of 21 tonnes and two large solar panels with a span of nearly 30 metres providing power. It had docking ports to receive visiting spacecraft and permit other modules to be added, increasing the workspace available and the range of experiments which could be undertaken.

One such module was Kvant, an astrophysics research module designed for experiments in materials processing which was attached to Mir in April 1987, although only after two docking failures which forced cosmonauts to make a space walk to clear an obstruction to the docking mechanism. Another was a module specialising in Earth observation.

Apart from the space station itself, the other big change for Chrétien was the length of the mission. Instead of a brief few days in orbit, he was due to conduct 160 hours of experiments over more than four weeks, a heavy schedule requiring detailed knowledge of each experiment prior to launch. Tognini, his back-up, needed not only to familiarise himself with the experiments but also learn how to fly Soviet spacecraft.

The launch was initially scheduled for 21 November 1988. Chrétien was due to ride into orbit with a two-man Soviet crew, who were flying to Mir to replace two other Soviet cosmonauts, Vladimir Titov and Musa Manarov, who had been on board for more than 300 days and who were heading for a record-breaking year aboard the space station.

All six men, including a doctor, Valery Polyakov, who had been on Mir since August, were then due to stay on board together for one month before Chrétien returned to Earth with Titov and Manarov.

A complication in the schedule arose when President François Mitterrand, due in the Soviet Union on an official visit that month, said he would like to attend the launch if it could be postponed for five days. With Titov and Manarov understandably anxious to return to Earth without further delay, the Soviet authorities decided simply to shorten Chrétien's flight to 25 days from the planned 30. This enabled Mitterrand to attend the launch while still getting Titov and Manarov back to Earth by late December.

There were strong contrasts between Chrétien's 1982 launch and his second journey into space more than six years later. Launches were now being shown live on Soviet television, something that would have been unthinkable in the secrecy-shrouded days of the 1970s and early 1980s, when even major space successes were reported after a delay of several hours, or even days. The change was due both to the more liberal political climate in the Soviet Union following the arrival of Mikhail Gorbachev as Kremlin leader, and to the growing confidence of Soviet space engineers in the reliability of their rockets. French television took a live feed of the launch broadcast, and French presenters were apparently caught unawares by another manifestation of perestroika on the TV screen – advertising breaks interrupting the Soviet launch coverage, with Russian companies paying around £10 000 each to sponsor the flight and win television air time for their products.

The Franco-Soviet mission, while capturing the headlines in France, was in fact rather a sideshow as far as the Soviet public was concerned. For less than two weeks earlier, on 15 November, the Soviet Union had successfully launched its first space shuttle, picturesquely called Buran (snowstorm). The pilotless craft made a fully-automatic two-orbit test flight before coming down to a perfect landing at Baikonour three-and-a-half hours later. The flight, although anticipated for several months, nevertheless came as dramatic confirmation of the Soviet Union's capabilities, not only to work at long-duration space flight using relatively traditional technology but also to follow the US into the more high-tech areas of space exploration and exploitation.

The white Concorde aircraft carrying President Mitterand and Soviet Foreign Minister Eduard Shevardnadze to view Chrétien's launch touched down on the same runway at Baikonour that Buran had used 11 days earlier.

There followed a meeting between Mitterrand and the crew who, wearing white suits, spoke to the French president from behind a glass screen. Chrétien, looking relaxed, gave an interview to French television news, playing down the tension of the impending launch.

> You have to remember that it only takes nine minutes to get into orbit, and those nine minutes pass very quickly. There are several things that have to be done, and you don't have much time to relax and enjoy yourself. . .[3]

Then the crew were driven out to the launch pad as the countdown entered its final stages.

It had been foggy earlier in the day and very cold, with temperatures down as low as minus 30 degrees Celsius. But then the fog lifted, giving a clear view of the rocket from the VIP observation area just 800 metres away. As night fell, the floodlights were turned on and at 18.50 Moscow time, mid-evening at Baikonour, the 310-tonne rocket soared away from the launchpad. It was an exceptionally clear night, and although the television camera lost the rocket quickly, observers on the ground were able to watch its trajectory for several minutes. Live television coverage from two cameras inside the cockpit showed the cosmonauts apparently

very calm, with mission commander Alexander Volkov impassively read-
ing flight documents.

> It was very impressive, because we were so close to the launchpad. The
> noise of the lift-off, and the plume of flame . . . it was quite blinding. And
> there were clouds which reflected the light and gave a spectacular
> backdrop.

Philippe Couillard, director of manned space flight at the French space
agency CNES, said after the launch.[4]

ABOARD MIR

The Soyuz spacecraft docked with the Mir space station two days later, to
an ecstatic welcome from the Soviet cosmonauts on board, two of whom
had been in space for more than 11 months. But while for Titov and
Manarov the arrival of the new Franco-Soviet mission marked the final
stage of their long stay on Mir, and perhaps a chance to relax a little, for
Chrétien it was the signal to throw himself into the very full programme of
scientific work that lay before him.

The highlight of the stay aboard Mir was a spacewalk by Chrétien and
Volkov, planned for 12 December but then brought forward three days to
permit time for a second walk if they failed to complete their programme.
It was the first time that a European had ever attempted a spacewalk, so
notching up another first for France within the European space community,
although Soviet and American astronauts had been doing them for more
than 20 years. As it happened, Chrétien did not simply become the first
European to leave a spaceship, he and Volkov also shattered the previous
duration record by staying outside Mir for six hours and ten minutes
instead of the three-and-a-half hours planned.

The reason for this was that they encountered major difficulties with
their main task, the deployment of a prototype structure designed for use
in future space stations as a support for aeriels, solar panels and similar
items. The structure, called ERA, had been built by Aérospatiale of
France and was a folded series of carbon-fibre tubes which, when opened
out, produced a complex grid four metres in diameter and one metre high.
Although it had more than 5000 components and 1500 joints, it weighed
just 44 kilos. Chrétien and Volkov installed it according to plan and linked
it by cable to the Mir control deck. But when flight engineer Sergei
Krikalyov pressed the button for it to deploy, nothing happened.

The cosmonauts tried jolting the space station to persuade ERA to
deploy, but without success, and Mir temporarily disappeared from contact
with the ground on the far side of the Earth, leaving teams of engineers at
the control centre puzzling over the problem. When contact resumed

Volkov laconically announced that the structure had been fully extended – throwing away his flight manual, he had resorted to the time-honoured method of kicking it.

Having tested its deployment, the structure was then jettisoned before the cosmonauts climbed back aboard Mir. Earlier in the walk they had also placed a rack of samples of paint and other materials on the outer wall of the space station, with the aim of leaving them exposed to cosmic radiation, micro-meteorites and the other hazards of the space environment for around six months before bringing them back inside and returning them to Earth for analysis.

Inside the space station Chrétien had a full schedule of experiments. He conducted more of the echocardiograph work initiated during his earlier flight and continued by Baudry on Discovery, using a specially-developed instrument called l'As de Coeur (ace of hearts). He also did experiments studying the effects of weightlessness on such diverse phenomena as posture control and body movement, calcium loss from bones and disruption to sleeping patterns and other bio-rhythms.

Non-medical experiments included measurements of gamma radiation inside the space station and a study of the effect of heavy ions on electrical components.

In a televised interview from Mir about a third of the way through the flight, Chrétien joked that he had not had much time to relax, before admitting that he had after all squeezed in a little sight-seeing, albeit of familiar Paris landmarks as he had sailed overhead.

> I saw the lights of the Eiffel Tower, it's really extraordinary to be able to see details like that. I saw the Eiffel Tower, the Champ de Mars, the black dot which was the Luxembourg Gardens. . .

And asked if he envied his two Soviet colleagues who had been in space for nearly a year, he replied

> Yes, in a way. They are installed here almost as if it was their dacha. They do their work with a great serenity and good humour. They are in an extraordinarily good state, so yes, I do envy them.[5]

Living conditions on Mir, although still cramped, were better than on previous Soviet stations, with each cosmonaut having the luxury of a private cabin. But Chrétien himself had contributed to making his flight more civilised than most. After his 1982 flight he had complained to the Soviet space authorities about the quality of food on which cosmonauts had to survive, traditionally unimaginative protein pastes squeezed from tubes. So in a bold gesture aimed partly at improving the cosmonauts' standard of living and partly at giving a little publicity to French cuisine, bids were invited from French food companies in late 1987 to supply cordon bleu meals for the Franco-Soviet space flight.

The competition was won by gourmet food company Comtesse du Barry and the dishes selected read like a menu from Maxim's, including such delicacies as duck with artichokes and rabbit with prunes. French gastronomy had to compromise with Soviet technology to the extent of putting the food in tins for the flight, but the dishes marked a big improvement on food served on previous missions.

So it was on a positive note, with scientists on the ground happy with the way Chrétien's exhaustive schedule of experiments had been conducted and the astronauts voicing appreciation at the improvement in their cuisine, that Titov, Manarov and Chrétien prepared to return to Earth. For the two Russians, the ride back was to mark the end of a record-breaking year in space.

It was supposed to be a routine descent and re-entry but there was a brief dramatic moment of concern when a computer overload on their Soyuz capsule forced them to spend an unscheduled extra three hours in orbit before initiating the re-entry burn. While the problem proved not to be serious, it recalled the predicament in which Soviet cosmonaut Vladimir Lyakhov and his Afghan colleague Abdul Ahad Mohmand had found themselves on a similar spacecraft just over three months earlier, when faulty computer commands caused two re-entry attempts to be aborted, prompting 'marooned cosmonaut' headlines in the Western press and predictions of a fiery fate before they landed safely nearly 24 hours late.

Chrétien, Titov and Manarov parachuted to a safe landing in Soviet central Asia on 21 December, bringing the space endurance record set by the two Russians to 365 days 22 hours and 39 minutes – just over a year or just under, depending on whether the fact that they were in space during a leap year was taken into the calculation.

The flight was seen as a great success both in France and the Soviet Union. The French space authorities began planning for the possibility of further space flights, while Soviet leader Gorbachev heaped praise on Chrétien during a Kremlin award ceremony for the cosmonauts held in January 1989, referring to the need for closer co-operation among nations in space due to the increasing complexity of space programmes. This repeated a theme that he had developed in a speech to the United Nations the previous month, while Chrétien was still in orbit, calling for international space co-operation, particularly in the area of environmental monitoring.

The only small note of dissent came in an article by Sergei Leskov published in the Soviet communist party youth newspaper *Komsomolskaya Pravda* a couple of days after the mission ended. This criticised the Soviet practice of giving free space flights to foreigners and called on the Soviet authorities to start charging realistic prices for Soyuz tickets. However, he also suggested that those foreign cosmonauts involved in useful experiments should pay less than those whose presence was largely a political gesture. Given the generous French space budget, even a policy change along the lines suggested seemed unlikely to deter further joint missions.

In fact, shortly afterwards the Soviet authorities did start offering space flights to foreign customers at a price. A Japanese television station expressed interest in broadcasting a breakfast show from Mir. British scientists, frustrated at the lack of government funding for their space activities, canvassed industry for financial support to send a British astronaut privately on a Soviet space mission. And Austria accepted an invitation to send a paying Austrian astronaut on an eight-day Soviet space flight in 1991. The Soviet authorities, grappling with massive budgetary problems and looking hard for ways of saving money in once-sacrosanct areas of spending such as defence and space, were beginning to realise that their hugely expensive space programme could begin to pay for itself, at least in a small way, and earn some hard currency.

13 Satellites and Probes in the 1990s

One of the fastest-growing areas of space research at the start of the 1990s was Earth observation, or remote sensing as it was also known. This covered a host of disciplines ranging from weather forecasting to geological surveying and crop monitoring. It also had major potential for tracking pollution and environmental hazards. Astronauts had already reported a dramatic deterioration in the clarity of the view of Earth from space in the decade between the mid-1970s and the mid-1980s, attributing this to atmospheric pollution. But a new generation of spacecraft being developed in Europe, the US and Japan and ready for launch in the early 1990s would be able to monitor such developments in a much more scientific way.

ENVIRONMENTAL MONITORING FROM SPACE

Environmental issues, after going through a period of being regarded as a minority concern, were beginning to force their way into the headlines and onto the political agenda in the developed world. The nuclear accidents at Three Mile Island and Chernobyl, the slash-and-burn deforestation of the Amazon, the crumbling of historic buildings in ancient cities such as Rome and Athens, the 1989 Alaska oil spill, the disappearance of a growing number of plant and animal species, the relentless spread of urban sprawl, the large-scale pollution of rivers and oceans and the steady destruction of the ozone layer all combined to create an atmosphere of environmental crisis, with dire warnings from pressure groups coupled with urgent calls for more information on the situation, and above all action.

The contribution of space science to monitoring the Earth's environment lay essentially in providing raw information from a new perspective, information that frequently was either unavailable from ground level or from conventional aerial surveillance, or simply too expensive or time-consuming to gather.

In August 1987 Sally Ride, America's first woman in space, presented a report to NASA, commissioned by Administrator James Fletcher, setting out the goals that she believed the agency should set itself. One of the key programmes that she recommended was a 'Mission to Planet Earth' which called for a comprehensive system of orbiting polar and geostationary platforms built by the US, Europe and Japan to closely monitor the Earth's environment and predict changes due either to natural processes or to human activity.

Mission to Planet Earth is an initiative to understand our home planet, how forces shape and affect its environment, how that environment is changing, and how those changes will affect us ... We currently lack the ability to foresee changes in the Earth System and their subsequent effects on the planet's physical, economic and social climate. But that could change; this initiative would revolutionize our ability to characterize our home planet and would be the first step toward developing predictive models of the global environment.

the report said.[1]

The Ride report proposed the launch of four scientific platforms into polar orbit, two provided by the US, one by ESA and one by the Japanese space agency NASDA, as well as five geostationary platforms, three built by the US, one by ESA and one by NASDA, to provide comprehensive real-time monitoring of the atmosphere, sea and vegetation. The Ride report, while only consultative, did provide a framework for work already being done in Europe on new remote sensing and meteorological satellite systems, holding out the possibility that they might function not simply in isolation but also as part of a grander global network.

ESA was working on a remote sensing satellite project, ERS-1 and its more advanced follow-up, ERS-2, which would for the first time produce all-weather high-resolution images by means of an Active Microwave Instrument (AMI) which could operate either as a wind scatterometer monitoring the direction and speed of the wind at the sea's surface, or as a synthetic aperture radar (SAR), scanning bands of ocean 100 kilometres wide. Because its equipment would operate in the microwave part of the spectrum, rather than the visual or infra-red used by the Spot and Landsat satellites, it would be able to obtain images through cloud cover, very useful both in polar regions for observation of ice packs and in equatorial regions for the study of deforestation. The role of the ERS was essentially scientific, although its value was also to test new observation systems and demonstrate the capabilities of European industry at the forefront of remote sensing technology.

We shouldn't underestimate the showroom effect that Spot has had for French technology ... ERS-1 should be an opportunity to show that Europe is present. You will have to wait until 1996 or 1997 to see any US equivalent

one senior ESA official commented.[2]

ERS-1 was designed to produce images with a 30-metre resolution, not quite as sharp as Spot's 10- to 20-metre resolution. However, its all-weather capability would mean ESA could guarantee around 20 000 images a year to clients from the satellite, delivered within three hours of being taken. Its orbit would cover the entire globe every 17 days on a dense-grid basis and every three days on a less dense grid. It was to study coastlines, polar ice behaviour and oceanic phenomena for scientific analysis, but also for use by industry. Offshore oil companies, fishing fleets and shipping groups were expected to benefit especially from improved reports on the weather, ocean currents, winds and ice movements.

The satellite would be able to measure the sea temperature to within 0.2 or 0.3 degrees Celsius and, by measuring the roughness of the ocean surface, gauge windspeeds between two and 20 metres per second with just a 10 per cent error rate. It would be able to detect oil pollution on sea and monitor phenomena such as deforestation and the spread of deserts on land. Nevertheless, at the pre-operational phase of the project, as in most other areas of space technology, officials conceded that while potential applications were numerous, customers for the information tended to be slow to come forward. 'I feel today that there is still more technology push than user application pull', one official commented.[3]

One use being developed experimentally for information such as that ERS-1 was intended to provide was that of preventing locust plagues in Africa and the Middle East. The United Nations Food and Agricultural Organisation (FAO), based in Rome, had been experimenting since the early 1980s with rainfall information gathered from Meteosat and vegetation monitoring based on US satellites to compile ten-day and monthly maps of African rainfall and vegetation changes. By analysing this data, likely locust breeding grounds could be pinpointed and preventative measures theoretically taken to eradicate the insects before they started swarming. In practice eradicating the locusts tended to prove more difficult than identifying their breeding grounds, but the project was still not properly operational by the end of the 1980s and it was hoped that improved data from more sophisticated satellites such as ERS could help in the battle against one of the more intractable African scourges.

ERS-1 was due to be launched in 1990, with ERS-2 taking over from it in orbit three years later. In addition the polar platform, approved at the Hague ministerial meeting and planned for launch in 1997, was intended to complement ERS data, with extensions to the polar platform foreseen for the year 2000 and beyond. And the second generation Meteosat satellites, providing enhanced data on such areas as atmospheric water vapour content and temperature profiles to improve weather forecasting, were due for launch from 1995 onwards.

The studies which we are currently conducting are only the basis for deciding what instruments the second-generation Meteosat and the polar platform should carry. The general aim, however, is to equip them with microwave instruments to plot profiles of the atmosphere. In effect the new meteorological satellites should be much more than simple observation points. They will be multimission centres for studying the physical characteristics of the atmosphere

one Italian industrialist associated with the Meteosat programme explained.[4]

While ESA was working on the ERS programme and Meteosat, France was engaged in a separate, more specialised project with NASA called Topex-Poseidon. It was intended to study ocean currents in detail by measuring wave heights to within an accuracy of as little as ten centimetres, orbiting the Earth at an altitude of 1330 kilometres and covering the entire surface of the globe every ten days. It would have two radar altimeters on board, one with a 200 kilo mass built by the US with proven technology and a smaller, more experimental one built by France weighing 25 kilos. The project had started as two separate programmes, the Topex programme being developed by NASA since 1980 and Poseidon being pursued by CNES since 1981.

When we realised in 1983 that we were basically trying to do the same things, it seemed a good idea to join forces . . . to build a better mission for less cost

one programme official said.[5]

Apart from each supplying an altimeter, NASA would provide the satellite bus and two tracking systems, needed to provide a precise reference for the satellite's position, without which the altimeter readings would be unintelligible. CNES would also provide a tracking system. The programme had grown out of an earlier satellite project called Seasat, which had been launched in 1978 and which functioned for three months before failing. While its life had been considerably shorter than planned, it had succeeded, along with the Nimbus 7 satellite launched around the same time, in whetting the appetite of the scientific community for more data on the oceans.

What these missions have done is to demonstrate unequivocally that space-borne techniques can play a major role in oceanography . . . if you don't know the circulation of the oceans, it's very difficult to know how it acts as a fly-wheel in the world climate system

another official associated with the programme commented.[6]

As with ERS-1, it was hoped that the Topex-Poseidon satellite could provide data which was not only of interest to the scientific community but also of commercial value to the fishing, shipping and oil industries. It was due for launch in 1992 on an Ariane rocket, with an anticipated operational life of three to five years.

ASTRONOMY FROM SPACE

While remote sensing was a key part of ESA's scientific programme up to the turn of the century, astronomical satellites also had an important role to play. One such was Hipparcos, a satellite designed to map the stars and named after the ancient Greek astronomer who set himself a similar task. While star maps had been compiled for centuries from Earth-bound observatories using increasingly sophisticated equipment, they all faced the same basic problem that measurements of the tiny amounts of light reaching Earth from the stars had to be taken through the distorting cushion of the atmosphere, making precise readings impossible. The Hipparcos satellite, launched in August 1989 and designed to operate for two and a half years, was intended to systematically scan 120 000 stars with five times the accuracy of ground observatories, measuring their distance from Earth and positions relative to each other. The £200 million-satellite was built by consortium led by Matra of France, with Matra responsible for the payload and Aeritalia building the platform and conducting the integration and testing work. The mission was dealt a significant blow, however, when the satellite's apogee motor failed to fire after launch, stranding it in the parking orbit where the rocket had left it and rendering it capable of gathering only some of the data that had been hoped for.

Another scientific mission planned by ESA for the 1990s was the Infrared Space Observatory (ISO), a satellite designed to study stars and other celestial objects in the infra-red band and to delve deep into the mystery of the origin of the universe.

Astronomers interested in studying the formation of stars need to be able to analyse the interstellar dust and gas clouds from which they emerge. Since they emit large quantities of infra-red radiation, an infra-red telescope can provide valuable information on such matters as the temperature of the dust clouds, the behaviour of their chemical components and the density of ions. Due to the fact that a large amount of infra-red radiation is blocked out by the Earth's atmoshere, however, a space-based telescope can provide data up to a thousand times more accurate than any ground observatory. As a result the ISO project was of potentially huge interest to the scientific community in offering a major step forward in furnishing knowledge about the formation of galaxies and ultimately the big bang, believed to have been at the origin of the universe some 15 billion years ago.

The satellite, due for launch in 1993, would be a highly unusual structure, since the infra-red telescope needed to be cooled to the extremely low temperature of minus 271 degrees Celsius, just two degrees above absolute zero, in order to work. As a result a central feature of the satellite was a large tank containing helium, which would cool the instruments as it slowly boiled off over a period of 18 months. 5.3 metres long and 2.5 metres wide, with a mass of 2.4 tonnes, ISO would be ESA's largest satellite to date. It would be equipped with a 60-centimetre-diameter telescope as well as cameras, spectrometers and other instruments developed by laboratories in West Germany, Britain and France. The enormous sensitivity of these instruments would permit ISO to take measurements across the immense void of space equivalent in sensitivity to measuring the temperature of a lighted cigarette on the moon from an observatory on Earth.

The satellite was being built by a consortium headed by Aérospatiale of France, which was responsible for construction and integration of the satellite, while MBB of West Germany had primary responsibility for the payload. It would go into an elliptical 24-hour orbit with a 70 000 kilometre apogee and 1000 kilometre perigee, with data transmitted through an ESA ground station at Villafranca, Spain. The total cost of the satellite, excluding launch, was estimated at Ffr 1.3 billion.

TELECOMMUNICATIONS AND DATA RELAY SATELLITES

In the field of telecommunications satellites, which was already well-developed by the late 1980s and one of the most commercially successful areas of space activity, several developments were under way. Eutelsat had ordered its second generation of television and telecommunication satellites, Eutelsat II, with higher capacity and power than the first generation vehicles already in space, and the first two launches were scheduled for 1990.

Separately ESA had been preparing for the launch in mid-1989 of Olympus, a 2.5 tonne telecommunications satellite intended to demonstrate the possibility of combining a wide range of services on a single spacecraft. With British Aerospace as prime contractor, it was intended to provide direct television broadcasts to home satellite dishes, intercontinental trunk telephone lines, trunk services within Europe, mobile communications, smaller-scale communications links for developing countries and specialised data-transmission services. With eight transponders providing two television channels as well as a range of other communication services, it was designed to have a five to ten year life. Like the OTS satellite launched in 1978, which had been the forerunner of a whole family of telecommunications satellites, ESA saw Olympus as an experimental spacecraft that would hopefully spawn many commercial spin-offs.

A rather different form of project was the Data Relay Satellite (DRS), which had become necessary due to the growing commitment of European countries to a host of future space activities ranging from manned space stations to an increasingly sophisticated fleet of satellites that required a correspondingly enhanced communications system with Earth. NASA already had data relay satellites in orbit to enable the shuttle to remain in permanent contact with mission control during all phases of its flight, apart from re-entry, avoiding the nail-biting communication blackouts that had occurred in the earlier days of the space programme when pioneering astronauts hurtling around the Earth in tiny capsules had disappeared at regular intervals round the far side of the globe into frustrating radio silence. The aim of the European DRS programme was to put two data relay satellites into orbit, one stationed over the Atlantic and one over the Indian Ocean in the second half of the 1990s, to provide communications for the Hermes mini-shuttle, the Columbus space station module and a range of satellites for upwards of 85 per cent of their orbits. The DRS programme was included in the overall space plan approved at the Hague meeting, although Britain argued that it could be a prime candidate for private funding by industry interested in using its facilities. The two satellites would avoid the need to install dozens of ground stations around the world to stay in touch with Europe's growing fleet of space vehicles.

The two DRS satellites were tentatively scheduled for launch in 1996, starting their operational lives by the end of that year.

Apart from the various civilian satellite programmes, European countries were involved in a number of projects to send probes out across the solar system.

DEEP SPACE PROBES

One of the most ambitious of these was the Cassini project to land a spacecraft on Saturn's largest moon, Titan. It was a moon that had fascinated astronomers since its discovery in the mid-seventeenth century. Deep-frozen, with a surface temperature of minus 179 degrees centigrade, washed by an ocean of liquid ethane and methane and swathed in a primeval atmosphere similar to that which enveloped the Earth four billion years ago, Titan offered a glimpse of an atmosphere in its early stages of development. While conditions did not exist on Titan for life to develop, given the extreme cold on the surface, the chemical reactions taking place within its atmosphere could shed light on similar processes that took place back on Earth before life first began to emerge.

The Cassini mission was to be a joint project between ESA and NASA, representing the European Space Agency's first venture into deep space since the Halley flypast in March 1986. The idea was to send two vehicles on a six and a half year odyssey to Saturn, with the launch scheduled for April 1996 and arrival in the Saturn system for 2002. They would be propelled in the latter stages of the flight by the powerful slingshot effect of Jupiter's gravitational field. The Cassini mother ship, built by NASA and

named after the seventeenth-century Italian astronomer who discovered
the gap in Saturn's rings, would swing into orbit around Saturn while the
European-built Huygens probe, named after Titan's Dutch discoverer,
would target in on Titan and descend down through the moon's atmosphere on an historic voyage of discovery. It would transmit data on the
moon's atmosphere back to Earth via the mother ship during its three-hour
180-kilometre parachute descent and, depending on the softness of the
landing, transmit a burst of data from Titan's surface before falling silent.

There was no doubt that Titan would be an extremely hostile environment, despite being the only moon in the solar system to have an
atmosphere. Its very high atmospheric pressure, one and a half times that
of Earth at its surface, and its extreme cold made it an unlikely candidate
for eventual colonisation by manned space missions. But the laboratory
that its atmosphere offered was exceptional. Composed of methane, nitro-
gen and hydrocarbons but lacking molecular oxygen, it closely resembled
the Earth's atmosphere shortly after it was formed and for this reason
presented a unique opportunity, not only to voyage out across the solar
system but to take a four billion year leap back in time to study an
atmosphere in its formative stages.

ESA's contribution to the mission was to build the probe that would
descend onto the surface of Titan. It was a British company, Marconi
Space Systems, which won an international competition to design the
probe, leading a consortium which included Dornier of West Germany,
Fluid Gravity of the United Kingdom, ETCA of Belgium, SEP of France
and Sener of Spain. While the probe's role would be to descend through
the atmosphere, the NASA vehicle in orbit around Saturn would make
more than 30 fly-pasts of Titan, swooping to within 1000 kilometres of the
surface to take readings with radar imaging remote sensing instruments.

A different type of co-operation between ESA and NASA, planned for
the mid-1990s, was designed to make detailed observations of the sun, and
the way in which variations in its behaviour influenced the Earth. There
were two separate elements to the programme, the Solar and Heliospheric
Observatory (SOHO) and a project involving four spacecraft, known as
CLUSTER, which would study plasma. The project was intended to take
further earlier work done by scientific satellites and NASA's Skylab.

SOHO would be put into orbit around the sun, travelling parallel with
the Earth at a distance of around 1.5 million kilometres from its home
planet. There it would be able to study the sun's surface and corona, as
well as the solar wind as it swept past Earth. The CLUSTER spacecraft,
orbiting the Earth in formation, would take co-ordinated measurements of
plasma and electro-magnetic fields. ESA would fund around two-thirds of
the combined projects, with NASA putting up the rest of the cash.
Instruments on the missions would be selected jointly, while European
firms would be given the contracts to build the spacecraft.

THE STRATEGIC DEFENSE INITIATIVE (SDI)

While transatlantic co-operation was developing steadily in the field of deep space research and solar physics, collaboration in the thornier area of military systems was much more problematical. President Reagan's Strategic Defense Initiative (SDI), the project announced in March 1983 to build a space-based shield against nuclear weapons and eventually make them obsolete, sparked massive controversy in the United States over its huge cost, estimated at around $25 billion, and in Europe over fears that it would destabilise arms negotiations and introduce a new element of instability in East-West relations.

In February 1985 US Defense Secretary Caspar Weinberger flew to Europe as part of a campaign to invite America's major allies to participate in the SDI programme, popularly known as Star Wars. The manner of the invitation, which came with a 60-day deadline for replies, caused some irritation, especially in West Germany where the issue was politically explosive. The three European countries most concerned by the invitation, Britain, France and West Germany, had substantially different perceptions of the problem. British Prime Minister Margaret Thatcher, having recently won re-election due in large part to the military victory against Argentina in the Falkland Islands, was able to safely ignore the fairly vocal opposition to Star Wars coming from anti-nuclear and left-wing groups at home and tell a joint session of the US Congress that Europe firmly supported SDI research.

France, while interested in the project and having less of a tradition of domestic political opposition to military programmes, had two main concerns. One was the effect that SDI research might have on destabilising East-West relations, and the other was its implications for France's own nuclear strike force. French defence policy for two decades had been based on its independent deterrent. A US programme designed to make nuclear weapons obsolete could have serious implications for the nation's defence strategy, especially as at the time it was in the process of undertaking an expensive modernisation of its strategic missile forces.

Chancellor Helmut Kohl's centre-right coalition government in West Germany was enthusiastic about the research opportunities offered by the programme but very wary of the political capital that the social democrats and environmentalists in the opposition would make of it. At a time when West Germany, a front-line NATO state, was trying to reduce the number of nuclear weapons on its own soil, it did not want a heightening of tension between Washington and Moscow. It was also concerned that if West German firms became involved in SDI, it should not become a one-way street, with German know-how being used by the Americans with no technological benefits flowing in the other direction. However, it was the geopolitical issues that dominated, with West German officials expressing

concern over whether SDI was aimed simply at eliminating the intercontinental missile threat to the US while having no effect on the nuclear threat to
Western Europe from Soviet short- and medium-range missiles and bombers.

Political opposition, especially in Britain and West Germany, was
countered with the argument that SDI would not only make the world a
safer place but that it would bring substantial research contracts to
European countries participating in the programme. In fact this was shown
to be an illusion at an early stage. Britain signed a memorandum of
understanding on SDI research in December 1985 and West Germany in
March 1986, in the expectation of headline-grabbing contracts to follow.
But by the end of that year very little had materialised. Out of total SDI
spending on research, development, testing and evaluation in FY 1986 of
$2.68 billion, Britain secured contracts worth just $28.9 million and West
German contracts worth $48.2 million, 1.0 and 1.8 per cent respectively of
the budget.[7]

Over subsequent years contracts increased as the SDI budget grew, but
the proportion of research spending finding its way across the Atlantic
remained relatively constant.

One reason for this was that US law, enshrined in the Bayh amendment
to the 1973 defence budget, stipulated that no Pentagon research and
development contracts could be given to foreign firms if a US company was
equally competent to carry out the work and willing to do so at lower cost.
In addition, not only were specific set-asides prohibited in the SDI budget
to fund foreign research, but Congress also laid down that 'US firms should
receive SDI contracts unless such awards would be likely to degrade
research results'.

This meant that European firms at best had to compete on a totally open
basis with US companies for Star Wars contracts, despite their relative lack
of experience with the negotiating mechanism, and in many circumstances
could face discrimination. The US government was also determined to
maintain complete rights over the use of SDI research results, and to
strictly control their dissemination abroad.

Leading European aerospace companies were naturally anxious to pick
up any SDI contracts coming their way, since they offered the opportunity
to participate in research in very high technology such as particle beams
and lasers, not only at no cost to themselves but also at no cost to their
national governments. But while some collaborative ventures did start up,
the combination of US reluctance to see too many tax dollars subsidising
foreign research, Washington's concern to keep SDI research firmly under
its own control for national security reasons and the political opposition in
Europe meant that SDI never became the major international collaborative effort that President Reagan hoped it might. With the US having
become the world's largest debtor nation by the time that George Bush
took over at the White House in January 1989, and with pressure from

Congress stronger than ever to hold down federal spending on what were regarded as grandiose projects, SDI appeared certain to face budgetary cutbacks in the 1990s that would severely dent the ambitions of its supporters.

Apart from SDI there were few space projects of a purely military nature in Europe, although some civilian projects were adaptable for military use and could count the armed forces among their customers. The Topex/ Poseidon oceanographic satellite, for example, while designed as a scientific research and applications orbiter, had an ability to accurately measure wave heights that could be used by the navy to detect surface turbulence caused by submarines manoeuvring down to a depth of about 200 metres. Perhaps more interestingly, the applications over land of such a system, capable of measuring contours with an accuracy of ten centimetres, had obvious potential in such areas as programming cruise missiles.

BRITISH AND FRENCH SATELLITE SURVEILLANCE PROJECTS

The two European powers with independent nuclear deterrents, Britain and France, were both interested in satellite surveillance, although France was rather further down the road. In 1987 a row broke out in Britain when the government obtained a court order preventing a left-wing journalist, Duncan Campbell, from talking or writing about a secret spy satellite project called Zircon. The BBC decided not to show a film Campbell had made for them about the project after police raided its offices and removed truckloads of material. And not only was the general public kept in the dark, but members of Parliament were also prevented from attending a private showing of the film. The court order was subsequently lifted, on condition that Campbell did not reveal any further information about the project.

France was meanwhile moving ahead rather more openly with its Helios spy-in-the-sky satellite programme, based on the development work already done on the Spot Earth observation programme but carried considerably further. Helios was a massive project, budgeted to cost more than \$1.25 billion and aiming to put four satellites into orbit, each with a three-year life, with a first launch planned for July 1993. Orbiting at an altitude of around 400 kilometres, it would be capable of taking pictures in the optical and infra-red bands with a resolution of just one metre, comparable with US and Soviet satellites and giving a much higher definition than Spot's 10- to 20-metre resolution. Helios would be capable of spotting military vehicles, missile silos and troop movements and be able to focus down even on tank tracks and ammunition crates. It would therefore be invaluable, not only for monitoring arms accords in which France's interests did not always coincide with Washington's, but also for helping the French military to keep an eye on events around the globe in support of France's interventionist foreign policy. Its military activities during the 1980s had ranged from battling rebels on the Pacific island of New Caledonia to exploding atom bombs beneath the ocean on Mururoa, patrolling the Gulf, stationing peace-keeping troops in Beirut and fighting the Libyans in Chad. With no indication that its global role in the 1990s would shrink, the need for a satellite-based observation system was obvious.

Helios would also carry on board equipment capable of detecting and analysing enemy radar installations, which would enable the military to improve the operational capability of the M-5 stealth missile being developed for France's new generation of nuclear submarines in the 1990s. While Britain appeared content to rely heavily on US satellite intelligence, as it had done during the Falklands conflict, France's determination to maintain its independence from Washington in foreign and defence policy made the Helios programme indispensable.

> The prospect of wide-ranging bargaining between Washington and Moscow on Euro-missiles encourages the government, as well as many defence experts in the socialist opposition, to believe that France needs its own observation satellites, in order to avoid depending on external sources of information to monitor disarmament accords

the influential daily *Le Monde* wrote in early 1987.[8]

The difference in emphasis placed by Britain and France on the need for military observation satellites was less a divergence in view on the usefulness of space technology than evidence of France's greater determination to pursue a foreign policy independent of Washington. Britain, even in the late 1980s, could count on greater access than France to sensitive US military satellite data, thanks to the much-vaunted special relationship between London and Washington. But France, with its superior technology, would be in a position by 1993 to obtain its own high-definition satellite pictures. Britain, in the absence of a comparable system, would be more dependent than ever on the goodwill of the US defence establishment.

14 Shuttles, Space Stations and Politics

When President Reagan issued an invitation to friendly nations in January 1984 to join the United States in building an international space station for peaceful scientific work, the European response was immediately enthusiastic. Reagan's publicly-stated goal of placing the station in orbit within ten years echoed John F. Kennedy's memorable pledge in the early 1960s to put a man on the moon by the end of the decade. While Reagan's project failed to have the same impact on the general public as Kennedy's battle cry, on both sides of the Atlantic it certainly galvanised the space industry, which was in need of long-term goals once the shuttle operations had become routine.

US INVITATION TO JOIN SPACE STATION PROJECT

NASA administrator James Beggs toured Europe in March 1984 formally presenting the invitation to the European Space Agency and interested governments, and West Germany and Italy jointly presented a proposal to the ESA council in June that year calling on Europe to supply a laboratory module 'Columbus' for the station. These two countries were particularly interested in the US invitation since their industries had played a key role in constructing and funding the Spacelab project – which at that stage had only had one flight – and were now looking for ways of building on that success. And the West German research and technology ministry (BMFT) was keen to fund further microgravity research by West German institutes, keeping the country in the forefront of European work in this area.

The following months saw several meetings between companies and organisations interested in taking part in the Columbus programme, and European ministers meeting in Rome in January 1985 formally endorsed preparatory work on the project 'as a significant part of an international space station programme'.[1]

The wording was important since there was some concern in Europe that the US, crippled by a budget deficit running at more than $100 billion a year, might be seeking to involve other nations in a very expensive space project as a way of defraying some of the costs, while retaining basic control of the space station. They were fears that subsequent negotiations showed to have at least some basis in fact.

The concept of the space station evolved slowly through a series of meetings between space officials from Europe and the US as well as Japan and Canada, which had also decided to join the programme. The basic design which emerged was one of a four-module station comprising two parallel rows of modules interlinked at the centre, with large metal structures stretching out from them supporting solar panels to power the complex. There would be three laboratory modules, one each designed and built separately by the US, Europe and Japan, and a fourth crew habitation module built by the US. Canada would build a large robot arm for external use. There would be docking ports for visiting spacecraft, and the possibility of associated scientific vehicles flying in close proximity to the complex, being monitored and serviced by astronauts. The total cost of developing and deploying the complex was put at more than $20 billion, with the US putting up 70 per cent of the cost. It was planned to launch the station into orbit in segments aboard 19 separate shuttle flights stretching into the mid-1990s.

From the start the European Columbus contribution to the space station was seen not only as an important technological development in its own right but also as a stepping stone to the construction of its own separate orbital facility in the more distant future.

> Europe's eventual goal is an autonomous space station which will provide European scientists and industrialists with a permanent platform for research and the processing of key materials. However, to attain this lofty goal, a gradual evolution is envisaged

ESA stated.[2]

Indeed Europe's interest in taking part in the international space station, later officially named 'Freedom' by President Reagan, was only part of a much wider programme being developed by ESA and the more enthusiastic member governments to give Europe total autonomy in manned and unmanned space activities by the end of the century. The Columbus programme was one of three pillars of this scheme, the other two being a radically redesigned Ariane rocket and a mini-space shuttle, Hermes, to give European astronauts independent access to space.

THE ARIANE-5 ROCKET PROGRAMME

While West Germany and Italy had shown the most enthusiasm for the Columbus project and were ultimately to put up most of the money, France was the clear driving force from the start behind both the Hermes mini-shuttle and the new-generation Ariane-5 rocket. France in any case had always been the major funder of the Ariane programme, paying more than 60 per cent of the bill for the rocket's early development, and most of the senior Arianespace officials and virtually all the launch technicians at Kourou were French. The whole programme, while officially European, had a totally French flavour, an image reinforced by the nationalistic tone of the coverage given in the French press to each successive Ariane launch.

While the earlier Ariane rockets had all essentially been derivatives of the Ariane-1 three-stage vehicle, with power being added through extended fuel tanks or small strap-on boosters, the Ariane-5 was a different rocket resembling much more the huge external tank and solid rocket boosters used to put the US shuttle into orbit. It was to be 50.5 metres high, smaller than the 58.4 metre Ariane-4 and little taller than the 49-metre Ariane-3 and much squatter in appearance than its pencil-thin predecessors.

The propulsion system would consist of one large main stage, powered by cryogenic fuel, and two giant solid-fuel strap-on boosters. Both the boosters and the main tank would be 25 metres in length, the tank 5.4 metres in diameter and the boosters 3.1 metres. The boosters, designed to give the rocket initial lift to break free of the Earth's gravitational pull, would each carry 170 tonnes of fuel and burn for just two minutes, generating 500 tonnes of thrust. The main stage, with 120 tonnes of fuel, would produce 100 tonnes of thrust over nine minutes. The configuration of the upper part of the launcher would depend on whether it was serving a manned or unmanned mission. In the case of a manned mission the mini-shuttle Hermes would sit on top of the main stage like a winged rocket, giving a clean aerodynamic sweep to the overall profile. In the case of an automatic satellite launch, the Hermes would be replaced by an upper casing containing up to three satellites, with either a cryogenic or storable-propellant propulsion system, depending on the size of the payload.

The overall lift-off weight of the Ariane-5 would be 725 tonnes, compared with 470 tonnes for the largest of the Ariane-4 rockets and just 240 tonnes for Ariane 3. However, it still remained dwarfed by the 2000-tonne lift-off weight of the US space shuttle rocket system.

Ariane-5 was designed to launch satellites weighing a total of 5.9 tonnes into geostationary transfer orbit, compared with Ariane-4's 4.2 tonne capability and Ariane-3's 2.6 tonne upper limit. In the Hermes configuration it could launch 21 tonnes into low Earth orbit.

The whole rationale of Ariane-5's design was that it should be capable of launching either astronauts or satellites, but, unlike the shuttle, not both at the same time. For ESA's philosophy was that it made little sense to risk astronauts' lives on satellite launch missions which could be accomplished more cheaply and efficiently using an automatic launcher, and that astronauts should be held in reserve for space missions where a manned presence was essential.

'Spaceflight is risky, therefore a crew should only be used when necessary', Ernst Messerschmid, who was a European astronaut aboard an October 1985 Spacelab mission, told delegates to a French space symposium in April 1988, making a plea for higher safety standards in the world space industry. He said that to date 14 astronauts out of a total of 201 had been killed, equal to seven per cent of the total, while three out of a total 120 manned missions had been completely lost with their crew, equal to 2.5 per cent. 'It's too high, we cannot accept this', he said.[3]

European space officials said that the design of Ariane-5 had been driven by a three-fold desire to increase launch capacity, cut launch costs and boost the rocket's reliability rate to 99 per cent from below 90 per cent in order to make it safe for astronauts. The technological challenge of improving the rocket's reliability was considerable, 'not just a figure of nine per cent (but] ... an order of magnitude', as one commented.[4]

THE HERMES MINI-SHUTTLE

Along with Ariane-5 and Columbus, the third pillar of the European Space Agency's strategy for achieving full autonomy in space was the Hermes space plane. This was a French concept developed by military jet manufacturer Avions Marcel Dassault and the state aerospace group Aérospatiale, an orbiting vehicle whose design underwent radical changes through the latter half of the 1980s as security requirements became stricter in the wake of the Challenger accident. Its early design called for an orbiter 15.5 metres long and 5.1 metres high, with a wingspan of 10.2 metres. Its wings were to be delta-shaped, and highly swept back, with upcurved wing-tips to optimise its aerodynamic profile.

It would have orbital manoeuvring engines at the rear, but like the shuttle would glide back to Earth with no possibility of using the engines for position adjustment. Unlike the shuttle its engines would not fire during lift-off, since it would be positioned at the top of the rocket rather than strapped shuttle-style to the side. It would weigh 17 tonnes and carry up to four astronauts for between one and four weeks in independent orbit, or as many as six astronauts for up to three months if docking with a space station. It would be able to carry up to 4.5 tonnes of freight in its cargo bay, such as supplies for a space station, spare parts for satellites or material for scientific experiments, but would not carry new satellites into orbit. Like the shuttle, the two halves of its cargo bay roof were designed to swing open once it reached space. The cargo bay would be five meters long and three meters in diameter, giving a volume of 35 cubic metres. Overall Hermes would be about half the size of the US shuttle.

Hermes would be launched aboard an Ariane-5 rocket from Kourou, French Guyana, and the flight would be monitored and directed from a space complex at Toulouse. The main landing runway would be at the Istres test flight centre near Marseilles, with a back-up runway at Kourou.

The orbiter would be flown back to Kourou after a mission, strapped to the top of an Airbus jet.

The general Hermes design had been developed by Dassault during the early 1980s, and a model exhibited by the company at the Paris air show in June 1985.

However, Aérospatiale, as the company which had built the Diamant rocket and led development work on the Ariane, was also anxious to be involved in any eventual shuttle project and had conducted studies of its own. As a result, as the French space agency CNES was deliberating during 1985 how to proceed, and with which industrial group in the driving seat, the rivalry between Dassault and Aérospatiale intensified. At this stage it was still essentially a French project, although European ministers meeting in Rome in January 1985 had noted 'with interest ... the French decision to undertake the Hermes manned spaceplane programme, with a view to including this programme as soon as feasible in the optional programmes of the (European Space) Agency'.[5]

On 18 October 1985 CNES announced its decision. Aérospatiale would be industrial prime contractor for the whole programme while Dassault would be made prime contractor for aeronautical aspects of the orbiter's development. CNES explained that Dassault's responsibilities would cover

all work necessary to ensure the success of flight in the atmosphere (aerodynamics, aerothermics, flight quality, general design of structure and aerodynamic forms, determination and supply of thermal protection).[6]

Aérospatiale would therefore be responsible for the overall design and integration of the project, including ensuring that it met CNES requirements for operational efficiency in orbit, while Dassault would be concerned with the aerodynamics of a plane which would have to survive the searing heat of re-entry, slowing from 28 000 kilometres per hour (Mach 25) to a landing speed of just 330 kph within the space of 35 minutes. 'Project leadership is divided between two companies who until now have been in hot competition with each other and who now must convert this into close collaboration', a French aerospace journal commented after the decision was announced.[7]

The two groups had worked together on aerospace projects before, notably the Concorde supersonic passenger airliner and the Mirage 2000 fighter jet, but had not previously collaborated on a space venture.

In March 1986, five months after CNES awarded the main contracts to Aérospatiale and Dassault and more than a year after the meeting of European ministers in Rome which had expressed interest in the project, France made its official proposal to the European Space Agency that Hermes should become a European project. The ESA governing council quickly approved the suggestion and a one-year preparatory programme was initiated, designed to run until a meeting of European ministers in November the following year could endorse the project.

A range of European companies became involved in the preparatory studies for Hermes subsystems. In West Germany Messerschmitt-Boel-kow-Blohm (MBB) took responsibility for the propulsion system, Dornier for the fuel cells and environmental control and life support systems and ANT Nachrichtentechnik for the orbiter's data acquisition and communications. In France, apart from the prime contractorship responsibilities of Aérospatiale and Dassault, Matra worked on the electronics, Dassault on the orbiter's heat shield and re-entry flight controls and Aérospatiale on the software. ETCA of Belgium handled the on-board power system, Fokker of the Netherlands the orbiter's external manipulator arm, Aeritalia of Italy its thermal control and Construcciones Aeronauticas SA (CASA) and SENER Ingenieria y Sistemas SA of Spain its airlock system.

Associated work concerning both ground infrastructure and equipment for use outside the orbiter while in space was being conducted by Dornier and British Aerospace.

France, with its industry taking the leading role in the orbiter's design and construction, was planning to pay around half of the programme costs, and expecting between 10 and 13 other European countries to also come in on the programme. CNES director-general Frédéric d'Allest felt confident enough in October 1985 to say that his agency had 'no serious worries about raising the other 50 per cent of the costs'.[8]

While French government and industry was unanimous in their enthusiasm for the project, however, other major West European countries were less convinced of its merits. Despite some interest being shown by the British space industry, Margaret Thatcher's conservative government in Britain was sceptical about the need for such a plane and horrified by its cost – estimated at Ffr 13.6 billion in early 1987, plus a further Ffr 1.5 billion for each flight, assuming two flights a year.

In West Germany the situation was more complex. With MBB and other German companies anxious to take part, Research and Technology Minister Heinz Riesenhuber was battling hard to win government guarantees of funding. But Finance Minister Gerhard Stoltenberg, with an eye on the budget deficit, took the position that no new funds could be made available for such projects and that therefore they would have to be paid for out of the existing space budget. French officials, frustrated at the delay, began to hint that, while they would rather have the Germans aboard, they could also build Hermes on their own. In early 1986 the French financial daily, *Les Echos*, reflecting the mood of the French media, commented that 'in no way can the future of the space plane be indefinitely suspended while Bonn makes up its mind'.[9]

If Hermes was experiencing some political difficulties during its definition phase, the same was certainly true of the space station project.

This was far more complicated, since not only was there the need to mould the interest being shown in various European countries in Columbus into a coherent industrial programme, but there were also conflicting signals coming from across the Atlantic about US commitment to the project.

US DEBATE ON THE SPACE STATION

NASA had been actively pursuing the space station project since the early 1980s, although with little support, and sometimes outright hostility, from other government departments. President Reagan's statement on space policy in July 1982, setting out the long-term goals of the US government, made no specific mention of a space station, referring simply in general terms to the need to pursue research both of astrophysical phenomena and of the Earth viewed from space.

The space station project had to compete for scarce funds with other short-term programmes, such as the construction of a fifth space shuttle, or more ambitious long-term plans such as the possible establishment of bases on the Moon. During late 1983, as NASA sought to win presidential approval for funding of a space station, the idea was being opposed by a host of government departments and agencies. These included the US Air Force and Defense Department, whose primary interests were in launching unmanned military payloads, and the Office of Management and Budget, which was nervous about the costs of a permanently manned station and was urging a more cautious, step-by-step approach involving perhaps an initially unmanned platform.

In the final weeks of the debate, as the US government was inching its way towards a decision on the space station, West German Research and Technology Minister Heinz Riesenhuber visited Washington and gave a speech at the Cosmos Club, attended by several senior State Department officials, in which he urged the building of a space station and made clear that Bonn would like to be part of the programme. While it was difficult to gauge the effect this had on the ultimate decision, the call, coming from one of Washington's most powerful allies, was undoubtedly well-timed from NASA's point of view. [10]

In early December President Reagan took the decision to authorise the construction of a space station as an international collaborative venture, and in his State of the Union message to Congress the following month he announced it publicly, making the space station initiative one of four key goals for his administration.

America is too great for small dreams ... Tonight I am directing NASA to develop a permanently manned space station and to do it within a decade. A space station will permit quantum leaps in our research in science, communications and in metals and life-saving medicines which can be manufactured only in space. We want our friends to help us meet these challenges and share in their benefits. NASA will invite other countries to participate so we can strengthen peace, build prosperity and expand freedom for all who share our goals

he said.[11]

Notwithstanding the ringing tones of the announcement, neither the budget problems nor the political opposition had gone away, and during Reagan's second term in office, as negotiations with Europe and Japan over the station made slow and painful progress, the timetable slipped and the costs escalated.

In September 1986, in a bid to halt cost overruns, NASA proposed shrinking the original ambitious space station to include perhaps just the two US modules at first, with the European and Japanese to be added later, and a much smaller infrastructure than originally planned to support external power sources and construction facilities.

RENEWED PENTAGON INTEREST

Then in December 1986 the Pentagon, which had strongly opposed the original decision to build a space station, decided that it would be in its interests to have right of access to the station, including the right to conduct military and SDI-related experiments, and asked NASA to delay further talks with Europe and Japan while it formalised its position.

Part of the reason for the change of heart was that the shuttle had been grounded at the start of the year, and it appeared likely that once flights resumed launch rates would be considerably below the 24 per year that had been envisaged during the heady pre-Challenger period of optimism, cutting back on the opportunities for shuttle-based military experiments.

The renewed interest of the Pentagon posed a dilemma for Europe, which had been under the impression that the space station would be for peaceful purposes. President Reagan, in his State of the Union address, had ambiguously said that the station was to 'strengthen peace' and 'expand freedom', phrases which did not appear to exclude military use. Now Europe needed urgently to clarify whether the Pentagon's involvement could significantly affect the primarily civilian character of the space station.

Against the background of doubts over military involvement, an even more serious row was brewing over who would ultimately manage the space station. The European Space Agency had accepted NASA's invitation to participate on the basis of equal partnership, and on the assumption that it would have final control over activities in the European module. However, by early 1987 the US government was proposing that while the station should be managed by a multilateral board including all interested parties, NASA should chair it and have the power to make decisions where a consensus could not be achieved. This amounted to giving NASA the right to override European and Japanese objections on use of the station. While ESA was prepared to concede to NASA the right unilaterally to manage the station in an emergency, it wanted all other decisions taken by consensus, and where no consensus existed, no decision at all to be taken.

Aside from the management issue, the US was also proposing not only that it should have the exclusive right to use its own laboratory module, but that Europe and Japan should be restricted to just 50 per cent use of their own modules, with the US having the right to the other 50 per cent.

By January 1987 talks between ESA and the US were on the verge of collapse, with deadlock on the key issue of station management and considerable uncertainty on the European side over the Pentagon's intentions. A two-day meeting was held in Washington on 11 and 12 February, attended by the US, Japan, Canada and the ESA countries, in an attempt to clear the air, and succeeded in at least keeping the negotiations open. A joint statement issued after the meeting said that 'all partners reaffirmed their intention to co-operate on the basis of genuine partnership', the final phrase clearly referring to the row over station management. Doubts over the military uses to which the space station might be put were glossed over, the statement saying simply that all sides 'confirmed that the space station will be developed and used for peaceful purposes'.[12]

EUROPEAN CONCERN OVER MILITARY USE OF THE SPACE STATION

In fact countries within the European Space Agency took differing views on the issue of military activity on the space station. France, whose space and defence industries were closely interconnected and which took pride in its independent nuclear role in the world, would have had little difficulty tolerating some degree of military activity on the space station. West Germany, which already had a heavy US military presence on its soil, would also probably have had few objections. But countries such as Switzerland, which jealously guarded its neutrality, or Sweden, which was a long-time critic of the arms race, were in a different position.

Roy Gibson, director-general of the British National Space Centre, put the debate into context in late February, pointing out the possibilities for compromise which existed while insisting Europe should not allow itself to be bull-dozed by Washington.

The use of the space station must be for peaceful purposes . . . however, it should be pointed out that the Department of Defense has been conducting peaceful scientific research in space for many years . . . In the end it is essential that the US realises and understands our views. We are in no mood to follow co-operation at any price.[13]

Cost forecasts for the station were continuing to mount, with the original estimate of $8 billion having risen to $14.5 billion by February 1987, expressed in 1984 dollars, or a staggering $21 billion in real terms. NASA was coming under mounting fire from Congress for failing to provide realistic cost forecasts, while Congressmen were bluntly expressing scepticism that the station would produce anything like a satisfactory return on investment. In late March NASA reached provisional agreement with the Administration to push ahead with a smaller version of the space station, costing $12.2 billion (in 1984 terms). This would include the European and Japanese modules as well as the two US modules and solar panels fixed to a large metallic boom. A second phase, at a cost of a further $3–4 billion, could be added later, including additional booms and a servicing facility.

With NASA having won temporary respite from its Congressional critics, and some of the disagreements between the US and Europe smoothed over at the Washington meeting in February, prospects for the space station appeared to be improving. Further talks between ESA and the US authorities were held in early May, and ESA director-general Reimar Luest gave a cautiously optimistic assessment afterwards, saying that while the management issue remained to be resolved, progress had been made on the question of military involvement.

It has been declared to us that it is the State Department which is leading the negotiations, not the Department of Defense, and that the station would be a civilian station for peaceful purposes

he said.[14]

European ministers had been due to meet in the Hague in June to discuss the whole issue of European participation in the space station, but the meeting was postponed until November to give ESA more time to resolve the outstanding questions. However, a meeting in Frascati, near Rome, in September made little headway, with Jean Arets, ESA's head of international affairs, commenting afterwards that neither side had moved and that significant differences of opinion still separated them. And in October a senior NASA official on a visit to Europe warned that the US was ready to build the station by itself if agreement could not be reached with its foreign partners on how it would be run. 'We have committed a plan to build a space station and . . . will go ahead whether we get the agreements or not', Richard Halpern, NASA director of space-station utilisation, said in London just a month before the ministerial meeting.[15]

FINANCIAL PROBLEMS IN EUROPE

If there was uncertainty over the nature of the projects to be discussed, there was also some uncertainty over the readiness of ESA member countries to produce the necessary finance. France had made the running, pledging to finance close to half the cost of both Hermes and Ariane-5 and a sizeable slice of Columbus. However, within the West German government there was considerable unease about the huge costs of the programmes on the table, estimated at close to $15 billion in total. Bonn was pledged to take a 22 per cent stake in Ariane-5 and a 38 per cent leading role in Columbus, which it saw as a natural follow-on to its pioneering work on Spacelab. But Hermes was more controversial, and as the ministerial meeting approached, West German officials were reported in the press as saying West Germany would welcome a postponement of any decision on the mini-shuttle for at least a couple of years. German industry was also divided on the question, with those companies which stood to win space contracts obviously favouring higher expenditure but other sectors of industry fearful that increased space spending could starve rival areas of research of state funds.

There was little debate in Italy where both government and industry were enthusiastic about the Columbus project, which had potentially major spin-offs in contracts for Aeritalia and other aerospace firms. The government had indicated that it was prepared to fund around a quarter of the Columbus programme, and smaller but significant slices of Ariane-5 and Hermes.

FUNDING ROW IN BRITAIN

In Britain a full-scale row had broken out over the whole question of public funding for space programmes. The British National Space Centre (BNSC), set up in 1984 to co-ordinate space policy, had been conducting a detailed review of Britain's £110 million annual space budget and had recommended to the government in a report that this should be virtually trebled to close to £300 million if British industry and universities were to derive optimum benefit from the British contribution to ESA. Britain's annual space budget was a quarter that of West Germany and one-sixth that of France, with 80 per cent of it going on ESA programmes. In 1986/ 87 Britain contributed £79 million to ESA, and the BNSC proposed boosting this to £200 million, at 1986 prices, by 1991, a move which would have raised Britain's share of the ESA budget to 14 per cent from 6 per cent.

The BNSC's plea for more cash came at a time of strict control on government spending, however, with those requesting more funds expected to show precisely where the returns would come. The experimental nature of many space activities, such as the Hermes and space station projects, and the enormous outlay involved over a period of years made it difficult to draw up the kind of cost-benefit analysis that the Treasury accountants wanted to see. And more eloquent calls for a leap of faith in space financing proved no more successful.

> In spite of being deprived of funding over the past decade or so, the UK still has a lot to offer in the next steps forward in space ... it would be fitting for us again to be able to play a major role in European and indeed world space. We have the competence, we have the experience and we have the necessary enthusiasm and faith – may we please have the money?

BNSC director general Roy Gibson said in a speech in London in 1987.[16]

The request fell on deaf ears, for in late July that year news leaked out that the government had decided to ignore the BNSC recommendations and freeze the space budget. In an indication of the lack of importance that the government gave to the subject, confirmation of this came not in a formal announcement but in a brief reply given by Prime Minister Margaret Thatcher to a question in the House of Commons on 23 July concerning the BNSC's application for higher space funding.

As she told members of Parliament,

> It is quite correct that we have not been able to find the considerable amount of extra expenditure that was requested ... We spend through the taxpayer some £4.5 billion on research and development. We are not able to find any more resources. Therefore it would mean a switch of resources from one research or technology development to another. My colleagues felt that they could not make that switch and therefore we shall continue our subscription to the European Space Agency, but at present we are not able to find more money.

And in a comment reflecting her government's free market philosophy she added 'I hope that the private sector, if it is interested in the results from such research, will come forward with considerable resources'.[17]

The brief announcement was sandwiched between a question on that morning's Cabinet meeting and one about value added tax on bus fares. There was no further parliamentary discussion of the matter before the summer recess a few days later.

It came as little surprise when less than two weeks later Roy Gibson resigned as head of the BNSC. Gibson, a former director-general of ESA, had been recruited by the government as a figure of authority in the space world to bring greater cohesion to British space policy. Taking over as BNSC's director-general in November 1985, he had orchestrated its review of budgetary planning and long-term space goals under the impression that the government would listen carefully to what he had to say. The blunt refusal to come up with any more money at all came as a severe blow. The final straw was a refusal at the end of July by the Department of Trade and Industry to fund the £6–9 million needed to keep various space research projects involving British companies alive until the November ESA ministerial meeting.

'Space chief quits after cash rebuff', bellowed *The Times* over an item spanning five columns of its front page on 5 August. And former Information Technology Minister Sir Geoffrey Pattie, who had been a lonely voice supporting space research in the Cabinet and who had been responsible for Gibson's appointment, was quoted in the *Independent* as warning that other ESA countries would see Gibson's departure as due to 'Britain's abdication from space ... an indication that short-term considerations must prevail at all costs'.[18]

ESA director-general Reimar Luest was saddened by news of the departure, but determined that it should not further delay action on Hermes, Columbus and Ariane-5. 'There must be decisions taken soon ... European industry cannot wait any longer', he commented the day news of Gibson's resignation hit the press.[19]

BRITISH OPPOSITION TO EUROPEAN SPACE PLANS

During the following weeks the British government's position was viewed with increasing impatience in continental Europe, where space officials expressed surprise at its lack of interest in the Hermes and Columbus programmes being proposed by ESA while making clear that European countries would press on without British funding if necessary. In late October, shortly before the Hague meeting, British Trade Minister Kenneth Clarke told the House of Commons that 'we have argued consistently that these plans are grandiose, not well targeted and do not provide for enough industrial and commercial participation'.[20]

The core of the government's argument was that each space project should be firmly examined on its merits, costed and studied to see what its prospects were for providing a commercial return. It was an argument that was hard to fault from an accounting point of view, but it lacked vision and historical perspective.

At the annual meeting of the International Astronautical Federation (IAF) in Brighton in October there was a great deal of criticism of Britain's position from delegates in the corridors and bars of the conference centre.

Then in early November, shortly before leaving for the Hague, Clarke laid his cards fully on the table in an article in *The Times* in which he attacked ESA's decision-making process, described the ambitious Columbus and Hermes plans as ill-considered and made a plea for a greater role for private industry in large space projects.

> Demands to give priority to achieving manned space flight by the year 2000 show that political objectives are uppermost in the minds of some ... I believe that greater emphasis should be placed on scientific and economic objectives. Unfortunately the European Space Agency has not so far helped governments to agree on the balance. I believe that it has simply piled up grandiose proposals, seeking to pursue every objective regardless of cost

Clarke wrote.[21]

He added that in his view ESA, which he had previously branded as a 'hugely expensive club', had become divorced from the market place and was weighed down by a decision-making process that lacked commercial and industrial discipline.

So it was that the battle lines were well defined before the ministers gathered on the morning of 9 November in the Hague, a bleak, chilly autumn day with the grey of the sky merging with the grey of the sea on the horizon. Feelings on the prospects for a clear-cut decision on Ariane-5, Columbus and Hermes were mixed, with continental European newspapers tending to take a more optimistic line while the British press was in introspective mood, critical of the government's position but divided on the way forward.

The Times, in a stern editorial on the morning the conference opened, called on the British government to join other European states in seizing the high ground in space exploration to avoid being left behind by the US, the Soviet Union and the growing Japanese space programme.

> A European venture will be taking the steps that could put its participants in closer command of the heavens. Britain will be standing on the sidelines, congratulating itself on its prudence. The full extent of this folly is only now becoming clear

it warned.[22]

The *Financial Times* also devoted an editorial the same day to the subject, taking the more measured line that the value of large space projects was difficult to assess but calling for more consistency in government financial support for the space sector to give private industry a framework in which to plan. It described as 'palpable nonsense' the British position that ESA could not afford the new programmes. 'The European Community's combined gross national products cou'd readily support a US-sized space programme if there were any political desire for one', it said. European space spending at the time amounted to barely 20 per cent of US expenditure on the sector.

> The more difficult question is whether a European failure to establish a strong position in space would cause a large loss of competitive advantage in the 21st century against the US and Japan or, on the contrary, whether a massive commitment of human and financial resources to space might not crowd out investment in other technologies which have a more direct bearing on industrial competitiveness

the *Financial Times* said.[23]

THE GERMAN FACTOR

While the British delegation entered the meeting making clear that its position had not changed, there had been significant movement in the German position which opened the door to a European agreement on financing all three projects, with or without British support. After several weeks of agonised debate Chancellor Helmut Kohl's centre-right coalition government had decided to adopt a more positive attitude to Hermes. On 6 November, three days before the Hague meeting, Riesenhuber was able to announce that in return for a cut in the total ESA spending programme of the order of 15 or 20 per cent, West Germany was prepared to put up a sizeable slice of the funding for Hermes, as well as its already-announced support for Columbus and Ariane-5. This would be conditional on a review of the Hermes programme after about three years to assess its technological feasibility, prior to putting it into full development.

The change in the German position from one of reluctance to become involved in Hermes funding to one of being prepared to put up money on condition of a tight review procedure was of critical importance, for the three programmes hung together. Without the West German commitment it was difficult to see how financing for Hermes could have been put together, unless France had been prepared to foot an even more substantial part of the bill. Without Hermes Europe would have been left totally dependent on the US for access to the Columbus space station module, and without Hermes the rationale for Ariane-5's flexible design would have collapsed. Thus Riesenhuber's announcement of the new German position on 6 November was a key factor in ensuring the success of the following week's meeting.

A second factor which undoubtedly helped the conference was the selection of Riesenhuber to chair the meeting. Conference delegates and journalists were all impressed by his good humour and diplomacy, which prevented open warfare from breaking out between the British and French delegations and kept a constructive atmosphere alive through two days of difficult negotiations.

THE HAGUE NEGOTIATIONS

Ironically, the British government's bluntness in public over its space policy in the weeks preceding the meeting not only left other delegations in no doubt over Britain's position, but appeared to have the effect of softening the impact of Clarke's speech to the meeting. Other ministers around the table knew precisely what he was going to say, and felt that Britain's voice counted for less once it was clear that it was not going to contribute anyway to the new programmes. Clarke, in his speech, criticised the Ariane-5 and Hermes programmes as a 'dash for European manned capability' and said that Europe was concentrating too much of its resources on getting into space, and not enough on achieving results once it was up there.

> The supporting infrastructure and transportation programmes are now, in the (ESA) Director General's words, at 'the centre of gravity of ESA's future activities,' whereas originally they were tailored to the needs of future applications . . . we should be careful to avoid delusions of grandeur. Instead what is needed is a cool, hard-headed examination of the options.[24]

Clarke's call for more attention to be given to the user programmes, rather than the means of space travel, won some support around the table. But he found himself in a minority of one in his rejection of all three programmes as they currently stood. The British position on Columbus was qualified by the uncertain state of negotiations with the US over utilisation of the space station, and he left the door open for British participation in the project at a future date. However, Britain's interest in Columbus was focussed not so much on the manned laboratory element as an unmanned polar platform, carrying a range of scientific payloads primarily for Earth observation, which British Aerospace was interested in building.

Clarke also stunned his fellow ministers by turning down ESA's request for a 5 per cent annual increase in its general science budget up to 1992, calling instead for a thorough review of this as well. He argued that the science budget had already had a 27 per cent increase in real terms in the three years to 1989, and did not require an immediate further raise. While Ariane-5, Hermes and Columbus were optional programmes to which ESA members were free to subscribe as they wished, the core science programme was a mandatory part of its budget, which all member states were obliged to support according to pre-fixed percentages. British refusal to endorse even this part seemed to take its isolation within ESA to still greater heights.

GO-AHEAD FOR ARIANE-5, HERMES AND COLUMBUS

As other ministers made their speeches, announcing the amounts that each country was prepared to put into the various programmes, it became clear that the funding, and the political will, existed to press ahead despite the British objections. At the end of the two-day meeting on the afternoon of Tuesday 10 November a resolution was passed, with Britain abstaining, which gave the go-ahead for development of Ariane-5, Hermes and Columbus and authorising a 5 per cent annual increase in the general science budget. It approved a development budget for Columbus of 3.7 billion accounting units (approximately $4.1 billion), for Hermes of 4.4 billion accounting units ($4.9 billion) and for Ariane-5 of 3.5 billion accounting units ($3.8 billion), with a review of the Hermes and Columbus programmes after three years. The Ariane-5 budget included three qualification flights, the Hermes budget two qualification flights and the Columbus budget the cost of launching the various elements of the programme. The mandatory science budget was to rise to 217 million accounting units by 1992 from a previously-agreed 177 million in 1989. A separate resolution, calling for further talks with the US on the space station project, was passed unanimously.

ESA director-general Reimar Luest was all smiles at the end of the meeting. 'I had not expected to get so much. We got the three projects started ... on Friday morning I had not expected that. The German position was very decisive', he said.[25]

In return for the approval of the three programmes, Luest had agreed to the West German request that a tight review of ESA's budget should be carried out, with the aim of cutting overall spending from the new agreed level by between 15 and 20 per cent. He described this target as possible to achieve without endangering the overall balance of the programmes, while not attempting to detail how it might be done.

Under the accords France and West Germany emerged as by far the largest financial backers of the new programmes, with Italy in third place and other countries a long way behind. France agreed to finance 45 per cent of the two projects it had initiated, Ariane-5 and Hermes, and 13.8 per cent of Columbus. West Germany, which was most interested in scientific research in orbit and a follow-on to the successful Spacelab programme, would finance 38 per cent of Columbus, by far the largest contribution, 30 per cent of Hermes and 22 per cent of Ariane-5.

Italy, which was also most interested in Columbus because of the possibilities it gave its industry to participate, took a 25 per cent stake while contributing 15 per cent to Ariane-5 and between 12 and 15 per cent to Hermes. Belgium agreed to fund between 5.0 and 6.4 per cent of the three projects and Spain between 3.0 per cent and 6.0 per cent. Austria, Sweden, Switzerland and Ireland all agreed to contribute to Ariane-5 but not to Columbus, mainly due to concern over the military use that the US might make of the space station. Norway agreed to contribute to Ariane-5 and Columbus, but not Hermes, while Switzerland and Austria came in on the Hermes funding. The Netherlands and Denmark said that they would contribute to all three projects, while only Britain of the 13 ESA nations made no commitment at all.

The result of all the commitments was that the Hermes programme, viewed before the conference as the least certain to win approval, actually emerged as oversubscribed, due mainly to the last-minute West German commitment to fund 30 per cent. Around 99 per cent of Ariane-5 funding was pledged, close enough to the total to make no difference. Only the Columbus project was left underfunded, with just over 90 per cent of the estimated total underwritten. This was due to the unfinished state of negotiations with the US over the space station, however, with Britain indicating that it could pay as much as 15 per cent of the cost if there was a satisfactory conclusion and a polar platform was built.

'I think that the spirit in which this conference was conducted can give rise to a new impetus in Europe ... what we have started here is extraordinarily promising', Riesenhuber said at a concluding press conference.[26]

French Industry Minister Alain Madelin was jubilant. He had come to The Hague to win endorsement from his European colleagues for years of effort and money invested by France in preparing the Ariane and Hermes projects and shaping a coherent manned space policy for Europe, and had achieved everything he had set out to do.

> I think this is money well spent. The space dream is within Europe's financial grasp, and it is unimaginable that Europe should be absent from the conquest of space ... to go into space you will no longer need a Soviet or American passport – a European passport will do

he beamed.[27]

When Kenneth Clarke emerged from the meeting to brief journalists, he encountered a barrage of unusually sceptical questions, some of them outright hostile. The press, often not very objective at the best of times, had shown little sympathy or understanding for the British point of view during the meeting and had no intention of changing heart now. Clarke was in buoyant mood and appeared to enjoy the rough and tumble of the questioning, which was still considerably more sedate than the average House of Commons debate to which he was well accustomed.

Was he happy with the way the meeting had gone? 'A perfectly satis-factory outcome, in that everyone is paying for what they want to pay for'.

Why was Britain opposed to Hermes? 'We think this is imitating the Americans 20 or 30 years after the Americans have already done it ... other members have gone off on what we regard as a frolic of their own. There's no market for Hermes, it's just a spacecraft which will take men into orbit.'

Why no British funds for Ariane-5? 'Man-rating, in our opinion, is going to make it far less competitive *vis-à-vis* its rivals.'

Why the refusal to increase the general science budget? 'Space science has done very well in the last three years ... it's not our highest priority.'

Why was Britain unenthusiastic about manned space flight in general? 'Its scientific and industrial advantages are far from certain.'

Did Britain want to participate further in ESA, given its reluctance to provide new funds? 'We remain active, and indeed leading, members of ESA ... nobody wants us to drop out ... When the word "European" is put before a project there is a lessening of common sense which would be applied to domestic policy.'

What was the British government going to do now? 'We are going to have discussions with our own industry. What we spend on space will depend on what proposals come forward.... The best thing is to be an intelligent user of space ... We're a passenger, and we'll go on the most cost-efficient launcher.'[28]

The mood in the corridors afterwards was generally conciliatory. ESA had the political mandate and the funding to press on with its new projects, and the feeling was that the door remained open to Britain to clamber aboard at a later stage. Clarke had not ruled out participation in Columbus, and no-one had suggested, at least in public, that Britain should leave the organisation.

We have to calm things down. Britain has a long and rich tradition in aerospace, the question is whether it is in danger of losing it. There is a lack of interest in space in Britain...

French space agency president Jacques-Louis Lyons told a group of journalists.[29]

The conference set target dates for the development of the three major programmes, calling for Ariane-5 to fly in 1995 and Hermes by 1998. Columbus elements were intended to start going into orbit in the mid-1990s, before the first Hermes flight, but the timing was dependent on the eventual US timetable for space station construction.

THE COLUMBUS PROJECT

The Columbus programme, as it emerged from the Hague conference, consisted not only of the main laboratory module attached to the US space station, referred to as the Attached Pressurised Module (APM). It also included three other autonomous elements, orbiting separately from the space station and performing a variety of scientific functions. One segment, called the Man-Tended Free Flyer (MTFF), was a semi-automatic laboratory for activities such as materials processing, life science experiments and work on fluid physics. It would comprise a pressurised module equivalent to two of the existing Spacelab modules, plus an unpressurised resources module and up to five tonnes of payload. It was designed to function automatically but require visits by astronauts, using Hermes, about once every six months to monitor and change experiments. Once every three to four years the laboratory would be brought to the international space station for an overhaul.

A second vehicle, called the Polar Platform, would be launched into a polar orbit carrying a range of experiments designed mainly for Earth observation. It would be able to carry up to 3.1 tonnes of payload and would operate fully automatically for around four years. At that point it could be serviced by a space shuttle and given a renewed lease of life. Britain was particularly interested in this aspect of the Columbus project, since it had a tradition of pre-eminence in the field of meteorology and remote sensing which the scientific community was anxious to preserve.

A fourth element in Columbus was a small automatic laboratory called Eureca (European Retrievable Carrier), built largely by MBB and Aeritalia and weighing 4.4 tonnes, which was intended to be launched from the shuttle, stay in orbit for about six months and then be retrieved by the shuttle and brought back to Earth, where the experiments could be analysed. Its first flight was scheduled for May 1991.

The Columbus programme, with its four separate elements, covered a range of automatic, semi-automatic and man-tended experimental options in space which were intended to explore the most efficient way of building up future research activities.

In the US a similar project to Europe's Man-Tended Free Flyer existed, but privately-funded, called the Industrial Space Facility (ISF). Developed by Space Industries Inc and Westinghouse Electric Co, it was designed to be launched in 1991 and become fully operational in 1992. It would be visited periodically by astronauts, but its primary function would be automatic. Ultimately it was planned that several ISF vehicles should be built, dependent on commercial demand. 'These are little flying factories out behind the major research facilities on the space station', ISF executive vice-president and former Spacelab astronaut Joseph Allen said during a 1988 visit to Europe to drum up interest in the project.[30]

The aim was partly to form a technological bridge between the scientific facilities available on Spacelab and the much more sophisticated international space station, and partly to provide production facilities in space separate from the laboratory areas dedicated to research. While

NASA had agreed to launch at least two of the ISF modules on the shuttle in the early 1990s, and the US government had said it would rent between 50 and 70 per cent of the space available on the first of the modules, the rest of the capacity would be available for rent to private industry.

RESUMPTION OF SPACE STATION TALKS

Following the Hague ministerial meeting ESA returned to the negotiating table with the US over the space station project. European ministers had adopted a resolution confirming their continued interest in co-operating with the US, but setting out four conditions. Firstly, ESA insisted on having responsibility over the European elements of the space station, while accepting NASA's overall co-ordinating role of the programme, and also insisted on having full control of the Columbus elements such as the Man-Tended Free Flyer and the polar platform not attached to the station. Secondly, it wanted clarification that the US intended to use the European parts of the space station for peaceful research only. Thirdly, it wanted a disputes procedure drawn up. And fourthly, it wanted a full legal regime safeguarding the interests of all space station users.

Talks resumed in Paris in mid-December 1987 and finally an outline compromise was reached in a further round of talks in Washington in early February 1988, when the diplomatic skills of Heinz Riesenhuber and US Secretary of State George Shultz finally succeeded in breaking the deadlock. Europe accepted the long-standing demand that the US should have control over the whole of the permanent space station structure, including the European Attached Pressurised Module (APM). But in return the US gave in to European demands that ESA should fully control both the Man-Tended Free Flyer and the polar platform, which would normally be in independent orbit but which would be serviced from time to time at the station. The defence issue was left fairly vague, with a general agreement that the station was for civilian purposes, but that the Pentagon had the right to conduct experiments provided they did not involve the testing of weapons.

Once the outline agreement had been reached, the formal approval of the various space authorities involved was not long in coming. And on 29 September 1988, nearly five years after President Reagan issued his invitation to work with the US on the space station, European officials put their signatures in Washington to an accord. Appropriately, given the traumatic, protracted nature of the negotiations, the signing ceremony came on the same day that the space shuttle Discovery climbed into orbit from Cape Canaveral, the first flight since the Challenger accident, putting America back in space after two and three-quarter years of soul-searching.

While a great deal of intellectual and emotional effort had gone into hammering out the space station agreement, teams of scientists, astronauts and administrators back in Europe had been trying to work out precisely what ESA should do with the Columbus facilities once they were in orbit.

ESA began a campaign to try to interest private industry in the facility, stressing that it was often necessary to design space experiments three or four years in advance and pointing out the risk of wasting valuable time once the station was operational if industry were not sufficiently prepared or informed. 'It would be a disaster to have it all flying up there like a monument and not knowing what to do with it', commented Fredrik Engstrom, head of ESA's space station directorate.[31]

PREPARATIONS FOR AN ASTRONAUT CORPS

At the same time ESA set up an astronaut office, headed by a Spaniard, Andres Ripoll, to select and train would-be astronauts for the Hermes and Columbus programmes. In the late 1980s there were just three official ESA astronauts – Ulf Merbold of West Germany and Wubbo Ockels of the Netherlands who had both flown aboard Spacelab, and Claude Nicollier of Switzerland, who had yet to make a space flight. The other Europeans who had been into space, Germans Ernst Messerschmid and Reinhard Furrer and Frenchmen Jean-Loup Chrétien and Patrick Baudry, had flown aboard flights sponsored by their national governments and were not officially part of the ESA squad.

ESA was planning to build the squad up to around 40-strong by the end of the century, of whom 12 would be needed as pilots to fly Hermes and the rest specialists to work on the various Columbus facilities.

One problem that ESA and NASA were addressing was the need to shorten the preparation time for experiments on the space station and associated platforms. It needed typically between three and four years to prepare an experiment to fly on Spacelab, plus analysis time afterwards. 'A scientist's working life is 25 years. If he gets one result every five years, I am not sure he will use space', Jean-Jacques Dordain, head of ESA's space station and platforms promotion and utilisation department, explained.[32]

The situation was already beginning to improve, however. With Spacelab the four-year preparation for a one-week space flight represented a ratio of 200 hours available on Earth for every one hour available during the flight. As a result it could be worth spending 200 hours in a laboratory trying to resolve a problem in order to avoid risking spending one hour solving it in space. With the Columbus programme it was envisaged that a two-year preparatory period would be needed for experiments which would then be conducted by astronauts spending three months in orbit, a ratio of ten hours of preparation time available for every hour in space.

Therefore with the evolution of space laboratories, the astronaut's time was becoming proportionately less precious. This also meant that it could make economic sense to use an astronaut to perform routine tasks associated with certain experiments, rather than construct machines specially to do the task. 'Sometimes it will be much cheaper to use a man for repetitive tasks', Wubbo Ockels observed in a somewhat clinical analysis of man's usefulness in space. 'He's a robot, he uses 100 watts of energy and he weighs less than 100 kilos.'[33]

In a much wider sense, the role of astronauts had been undergoing a gradual change since the start of the space age. The early Gemini and Mercury space shots were essentially test flights, similar to putting a new fighter aircraft through its paces, and required experienced air force pilots to do the job. Science pilots started to fly aboard the Skylab programme in the early 1970s, but even with the arrival of the shuttle ten years later scientists were still outnumbered aboard the orbiter by crew whose primary mission was to fly it. The first flight of the fully-automatic Soviet shuttle signalled a quantum leap in space technology since, although the Soviet authorities were planning to use pilots for manned missions, the ratio of pilots to scientists on board could be substantially reduced and most of the crew of up to ten could concentrate on research jobs. In the space of 20 years missions had matured from concentrating simply on getting up into space and coming back again to focussing increasingly on accomplishing research or exploration once in orbit.

HERMES DESIGN CHANGES POST-CHALLENGER

In this respect modifications to the Hermes mini-shuttle programme, enforced in the wake of the Challenger accident, represented an inevitable but unfortunate step backwards. The initial Hermes design had called for a vehicle taking up to four crew when functioning as an orbital laboratory or satellite repair vehicle, and up to six when shuttling crews to and from a space station. Of the total crew, only two would be pilots, aboard primarily to fly the vehicle, and the rest would be scientists. Following Challenger, however, the perceived pressure of public opinion was such that it became necessary radically to redesign Hermes in order to make the cockpit section ejectable in case of an emergency. This resulted in a much smaller cockpit, capable of taking only three astronauts instead of the maximum of six originally envisaged. This was not simply a reduction in numbers, but also represented a major shift in the composition of the crew, since at least two of the three, and possibly all three, would need training in flying the Hermes, in addition to any scientific qualifications for space station work. The vision of scientist passengers, unskilled in flying the space plane, was dealt a severe blow.

I believe that three fully-qualified crew members will be necessary. I have difficulty in envisaging one pilot, one engineer and one scientist passenger who is not familiar with the Hermes system. Therefore I believe any passenger would have to be a fourth crew member . . . I don't know how we would do this

Claude Nicollier, a member of ESA's astronaut squad, said.[34]

Hermes' payload capacity also had to be reduced to three tonnes, 33 per cent less that the 4.5 tonnes originally foreseen and only 10 per cent of the US shuttle's capacity. However, cargo would now be carried in a pressurised compartment, instead of the unpressurised one envisaged in the original design. Roof flaps would no longer open out to expose the cargo bay to space, as in the US shuttle. Instead access to the cargo bay would be through an enclosed passage way.

The vehicle would be smaller and heavier than originally foreseen, its length being cut to 14.7 metres from an original 17.9 metres and its height falling to 3.2 metres from 5.1 metres. Wingspan at ten metres would be little changed from the earlier projection of 10.2 metres, however.

There had been considerable disagreement in France, where most of the design work was being done, about the desirability of making the cabin ejectable. While the French government and the space authorities, who were responsible for funding the project, realised that public opinion probably would not tolerate paying billions of dollars for a space shuttle without making every effort to avoid a repeat of the Challenger accident, some of the astronauts, including Hermes chief test pilot Patrick Baudry, opposed the move. Astronauts who had spent many years as test pilots were used to taking risks as part of their professional lives, and felt that the redesign of the Hermes to accommodate an ejectable cabin was compromising part of the object of building it in the first place. Halving the maximum crew size to three from six certainly represented a major alteration in a vehicle which was already much smaller than its US counterpart.

While there was never any real doubt in France that Hermes should be built, the West German government was only won round fairly late in the negotiating round leading up to the Hague conference. And while the British government remained resolutely opposed to it on principle, arguing against the need for Europe to invest in a manned space programme of any description, even those supporters of higher space spending in Britain had severe doubts about Hermes, partly because its small size restricted its uses and partly because it represented little technological advance over the US shuttle, being still reliant on an expendable rocket to put it in space.

'Hermes is a luxury. It has no future', Roy Gibson, former director-general of ESA, told a British Parliamentary committee on space in October 1987, two months after he had resigned his post as director-general of the British National Space Centre (BNSC) in protest at the lack of space funding by the British government.

> It would allow you to take two men and a ham sandwich into space from the time that it is ready, which is something like 1997 or 1998, until something else comes along, which is likely to be 2005....One has to ask – is it worthwhile for that period? In my view it is not worthwhile.[35]

SHUTTLES AND SPACE PLANES

Gibson's objections to Hermes went right to the heart of another debate, which had scarcely been mentioned at the Hague meeting but which had a direct bearing on Hermes' future, namely whether shuttles riding into space on expendable rockets had a significant role to play beyond the year 2000, or whether a new generation of fully re-usable space planes could be developed sufficiently rapidly to make investment in Hermes-type vehicles unnecessary.

In this debate the key factor was cost. Launch costs on expendable systems were high precisely because so much was thrown away during each flight. Launch costs would therefore only cease to be prohibitive for large-scale industrial development of space if fully-reusable launch systems were developed, the argument went.

> The show stopper in the whole affair is the cost of launching ... I think developing refinements of the present launch system, which goes up vertically, are strictly limited and we will never get the percentage decrease of cost unless we change, fundamentally, the way in which we are approaching launching

Gibson told the same parliamentary committee while still BNSC director-general.[36]

By the late 1980s scientists and engineers in several countries were working on futuristic designs for hypersonic space planes, designed to take off and land from airport runways. Some would carry passengers on hypersonic intercontinental hops through low Earth orbit, reaching speeds of up to 7000 miles per hour, 12 times the speed of sound, and reducing journeys such as London-Sydney or Los Angeles-Tokyo to a couple of hours. Others would carry astronauts into orbit to carry out research work on space stations, while still others would be unmanned and operate as cheap satellite launch vehicles.

Most of the money available for research into the space plane concept was in the United States, where several designs were being explored. NASA was co-ordinating the various projects within the framework of the National Aerospace Plane (NASP) programme, which aimed to build a prototype, dubbed the X-30, by 1995 to demonstrate and test the technology needed.

The prototype's name appropriately recalled the X-15 rocket plane flights of the early 1960s, when NASA was experimenting with high-altitude hypersonic flight and was just beginning to put astronauts into orbit. There were three of these rocket-planes built, and they were slung from the wings of giant B-52 bombers and launched at an altitude of 45 000 feet, after which a rocket engine fired to take them up to over 50 miles high. Dozens of flights were made, conducting valuable research into aerodynamics. After the discontinuation of the X-15 programme NASA gained further experience of hypersonic flight during the shuttle programme.

As with the space station and the Star Wars project before it, it was President Reagan who provided the political impetus to get the programme under way, proposing the construction of a hypersonic plane by the year 2000. But Reagan's announcement was long on dreams and short on detail, and it was left to industry and government between them to come up with credible computer drawings of what the craft might look like. Cost estimates were based largely on guesswork, with the Pentagon's original forecast that research and development for the project would cost just $3 billion questioned by some in the industry as wildly over-optimistic. In an article published in the Massachusetts Institute of Technology's *Technology Review* a graduate student in aeronautics and astronautics, who had worked at Rockwell International analysing the cost of the aerospace plane project, cast severe doubt both on official cost estimates and on the performance claims being made for the new vehicle.

> Much of this talk is hypersonic hyperbole. The technological challenges have been downplayed, the development costs are grossly under-estimated and the utility of the aircraft is vastly exaggerated. This vehicle would be the most complex aircraft ever attempted, and its true development cost is likely to be six times the official estimate.[37]

By mid-1987 NASA was referring to the aerospace plane project both for cheap access to orbit and for intercontinental travel. And industry, adding up the cost for itself, was concluding that it would be so high that certainly no one company, and probably no single country, could afford to develop such a plane by itself. McDonnell Douglas Corp senior vice-president Lou Harrington appealed just before the Paris air show in June 1987 for the world's major aerospace manufacturers to club together to work jointly on the project.

> This aircraft programme offers us the opportunity of major collaboration ... the key to unlocking the door of collaboration lies in the high cost of developing this aircraft, which some estimates place as high as $20 billion ... that's far more than any single company or some countries can afford.[38]

The formidable technological hurdles to be overcome in building the plane included designing a propulsion system that could cope with a range of speeds all the way up to Mach 25 and at the same time developing very strong composite materials capable of dealing with the immense temperature variations encountered during the climb to orbital velocity, and the subsequent descent.

While US industry grappled with the problem, rival firms on the far side of the Atlantic were anxious not to be left behind. The two projects being most actively developed were in Britain and West Germany, while French aerospace engineers were fully occupied with their work on the Hermes programme. The British project was called Hotol and the German Saenger, and while the problems that they faced were in many respects similar, there were significant differences in approach.

THE SAENGER PROJECT

Saenger was conceived as a two-stage craft, a sleek hypersonic plane the size of a Boeing 747 designed to take off and land at conventional airports, with an orbital vehicle called Horus (Horizontal Upper Stage) perched on top of it. Design work was being done by Messerschmitt-Boelkow-Blohm (MBB) and the project had been adopted by the West German Research and Technology Ministry as a reference concept for a national hypersonic programme. The lower stage would have six air-breathing turbo ramjet engines, taking it up to a cruising speed of around Mach 4.4 and close to Mach 7 just before the separation of the upper stage at a height of around 35 kilometres. Horus, which would resemble a more streamlined version of Hermes, would have two liquid hydrogen and oxygen engines to power it up to low Earth orbit at an altitude of around 500 kilometres. 27.6 metres long and weighing 22.2 tonnes, it would be designed to carry between two and six astronauts and a payload of two to four tonnes into space. Its primary role would be servicing and supplying space stations and transporting relief crews.

A second version of the upper stage, called Cargus, would be essentially an unmanned tube, 30 metres long and weighing six tonnes, carrying between five and 15 tonnes of cargo into low Earth orbit.

The lower stage could also be developed as a hypersonic passenger plane, carrying around 230 people at a cruising speed of Mach 4.4 with a range of 11 000 kilometres. This would give it more than double the passenger capacity of the Anglo-French supersonic Concorde, double the speed and also well over twice the range.

Comparisons with Concorde are inevitable, and enthusiasts of hypersonic transport are well aware that while Concorde, developed in the 1960s, was a stunning technological success for its time, it was also a financial disaster, with the British and French governments never coming close to recouping their multi-billion dollar investment costs.

If hypersonic transport is to be successfully developed in the future, it will have to overcome many of the objections that effectively killed Concorde as a commercial proposition. Concorde's limited range of around 5000 kilometres meant that it was capable of flying trans-atlantic routes such as London-New York or Dakar-Rio de Janeiro, but was unable to operate on the potentially lucrative Pacific routes linking the West Coast of the United States to Japan, the Far East and Australia. From Europe many of its potential routes, such as those across Asia to Australia or linking Europe with South Africa, were over land for much of the way, exposing the plane to opposition from countries along its flight-path not only on noise grounds but also because of fears that its sonic boom could damage buildings. Landing rights proved difficult to negotiate due to the plane's very high noise level, while environmentalists protested that cruising at above 50 000 feet, much higher than normal aircraft, it could damage the ozone layer. Landing rights were conceded at some US cities, notably New York, Washington and Miami, but the ban on it flying supersonically over the United States prevented it from developing routes between Europe and the West Coast. Its market was restricted to business travellers and wealthy private individuals on the few routes that were negotiated for it, as well as people prepared to pay large sums of money for exclusive charter flights to exotic destinations. French President François Mitterrand also took to using the plane for visits abroad, partly to impress his hosts with French technology and partly to cut the flying time to the more far-flung outposts of the Francophone world.

As a result, any political decision on the future of hypersonic transport must inevitably take into account not only the design challenges of the aircraft but also the commercial experience of Concorde. The German design for a hypersonic passenger plane goes some way to answering the criticism, since its extended range would open up many more route possibilities. The fact that much of its flight would be at extremely high altitude would also overcome many objections about overflying, although concerns over noise level and safety will inevitably spark protests from people living near airports where it is to operate.

The Saenger project was named after German space engineer Eugen Saenger, who first drew up designs for a two-stage rocket plane back in the early 1940s. This was at the same time as Werner von Braun was working on the V-2 rockets, but while the V-2s were aimed at London, Saenger was more interested in levelling New York. His idea was a plane travelling more than 17 000 kilometres an hour which would be capable of dropping up to one tonne of dynamite on Manhattan. Needless to say the plane was never built, but the designs remained and form the crude basis for the current project. In late 1986, when official interest in the Saenger programme was being aroused in Bonn, development costs over about 20 years were estimated at $12 billion. German interest in the Hermes project was partly explained by the fact that much of the work being done on it was directly applicable to the Horus shuttle vehicle in the Saenger programme.

THE HOTOL PROJECT

The British hypersonic plane project, Hotol, was rather different. Invented by Alan Bond of the British Atomic Energy Authority, it was conceived simply as a single-stage space plane that would take off horizontally and climb into orbit using revolutionary air-breathing engines that would extract oxygen from the atmosphere during the ascent phase before converting to liquid oxygen in space. Its air-breathing phase would take it to a speed of Mach 5 at an altitude of 26 kilometres before switching to rocket power for the climb to 90 kilometres and the swing into elliptical orbit. At the apogee of the orbit, around 300 kilometres up, orbital manoeuvring engines would push the plane into a circular low Earth orbit. For re-entry, thrusters would turn the plane tail-first, lining it up for a steep descent that would take temperatures on the nose and leading edges of the wing as high as 1500 degrees Celsius before coming into land on a normal runway at around 270 kilometres per hour.

The plane, rather than taking off on its own wheels, would be catapulted into the air from a runway trolley. This was because, with a take-off weight of 196 tonnes – much of it due to fuel on board – and a landing weight of just 34 tonnes, it would need a much heavier-duty undercarriage for take-off than landing. By launching it from a trolley, considerable weight could be saved on the landing gear.

The plane was principally designed to operate fully automatically and unmanned, simply taking satellites into orbit and launching them before returning to Earth. As a fully re-usable system, it was calculated that it would cut the cost of launching payloads by a factor of five, reducing the cost per kilo of launching a satellite to around $600 a kilo from close to $3000 on the shuttle.

The automatic version, with a cargo hold 7.5 metres long and 4.6 metres in diameter, would be able to launch payloads of between seven and 11 tonnes, and bring a similar mass back down to Earth. An alternative, pressurised version of Hotol would be designed to take between two and four astronauts into low Earth orbit for missions of between five and ten days.

British officials briefed the European Space Agency on the project in 1985, encountering interest but also some criticism because of a refusal to give details of the air-breathing engines, which were classified by the Ministry of Defence as secret material. Since the entire project depended on the feasibility of constructing such engines, it was difficult for other European nations to develop firm opinions on the project. The following year Sir Geoffrey Pattie, Britain's Minister for Information Technology, briefed other European ministers on the Hotol proposal, giving rather more detail but still refusing to reveal the secret of the engines. A document he presented to ministers, and ESA director-general Reimar Luest, estimated the cost of developing Hotol as around £5 billion, comparable at the time to the total estimated development costs for Hermes and Ariane-5.

The initial development phase was proposed to last until around 1991 and the subsequent development and manufacture phase until the end of 1997, followed by four years of flight trials. On this timetable Hotol could be operational by 2001, making a considerably more advanced space technology than Hermes available just three or four years later.

BRITISH FUNDING PROBLEMS WITH HOTOL

While the government was showing interest in putting up some initial money for Hotol, it was British Aerospace and Rolls-Royce which were making the running on the project, developing the designs, conducting the research and lobbying hard for funds. The cost of the proof-of-concept phase, during which engineers were simply trying to determine whether the design had a future and whether further investment should be made in it or not, was relatively cheap and did get financial support from state funds, although only to the tune of around £1.5 million. But during the review of science spending being conducted during 1987, when the British space budget was effectively frozen and Prime Minister Margaret Thatcher made clear that no new funds would be available, much of the impetus behind the Hotol project vanished. British Aerospace became increasingly convinced that the project was viable, but lacked the funds – estimated at around £6 million for the next stage – to push ahead with its development. European countries, perceiving the lack of political will in Britain to press for wider ESA funding for the programme and faced with intense French pressure to commit scarce resources instead to Ariane-5 and Hermes, saw no reason to take the initiative on a high-risk project on which they still had scant details and which just a few years earlier would have been regarded as pure science fiction.

> We now have a most bizarre state of affairs. Having proved the concept, we have left our companies high and dry at the very moment when the drive to find collaborative partners was about to happen

Pattie wrote after the government's refusal to provide further state investment in Hotol.[39]

The government took the view that foreign industrial partners needed to be found for the development of Hotol, but was reluctant in the meantime to fund British research. At the same time its refusal to join the Ariane-5, Hermes and Columbus projects at the Hague meeting and its negative stance on the increase in ESA's space science budget had created the impression in Europe that Britain was becoming an unreliable partner for the necessarily large investments involved in new space programmes.

> I believe had we been good partners in Europe we could have looked in the fullness of time to willing cooperation by European countries in the Hotol programme. I think there is an enormous question mark now against the position of the United Kingdom as a good collaborative partner in ventures...

a senior British Aerospace executive commented in late 1987.[40]

Exploratory talks were held between British and West German industry on their respective hypersonic plane projects, but they showed few early signs of bearing fruit. A major problem remained British reluctance to declassify the Rolls-Royce RB.545 engines which would power the Hotol, since refusal to discuss details of the air-breathing technology meant that it was difficult to take discussions with potential foreign partners beyond a relatively superficial stage.

By 1988 British Aerospace, despite the disappointments over government funding, was still optimistically pressing ahead with the project, seeking private finance to keep research work on track. Materials suppliers, who could have an interest in the high-technology research necessary to define the carbon-carbon and other lightweight heat-resistant components integral to the Hotol design, were one potential source of funds. But however the initial research phase was kept alive, it was clear that full development and production could never go ahead without both a high level of state funding and considerable involvement by other European countries. The British National Space Centre, in mid-1987, was envisaging ultimate British participation in the Hotol project at around the 25 per cent level, with not only other European nations participating in the funding and development but possibly also the Americans and Japanese as well. Whatever the merits of the British government's position, there was little question that diplomatically it had shot itself in the foot by adopting such an openly critical attitude to ESA programmes at the same time as encouraging British industry to seek foreign participation in Hotol.

Sir Geoffrey Pattie, in his testimony to the Parliamentary committee on space in late 1987, was particularly critical of the British government's attitude to Hotol funding, arguing that it appeared to be in danger of becoming the latest in a long line of technological projects that Britain had initiated and then abandoned, only for other countries to pursue and commercialise. The cancellation of the Blue Streak rocket project in the

1960s had cast a long shadow over the British space community, while longer memories recalled Frank Whittle's vain struggle to interest the British government in funding the development of the jet engine.

> There are plenty of people around who were actually involved in Blue Streak and who have an awful déjà vu sense about what is going on now. Certainly the Japanese and Americans would like very much indeed to discover how Hotol works, and I just hope that we are not looking at yet another system which Britain invents, starts to have an idea about and then the world says 'Yes, that is the way to have a reusable system'. . . . A horizontally launched reusable system is very likely to be what people will be using in the early part of the next century. Someone may then say in a small footnote, 'Yes, I think it was first actually developed in Britain' and it will all be done by Mitsubishi by then. Does that matter? I happen to think . . . that it does.[41]

In late 1988 the Carroll Group, a private property and investment group which had a Defence Ministry contract to develop land owned by the Royal Aircraft Establishment at Farnborough, Hampshire, indicated that it could be interested in putting money into the Hotol project. But in general private industry did not show much readiness or ability to come up with cash. With the government, British Aerospace and Rolls-Royce having spent an initial sum of around £3 million on research, the programme was looking for considerably increased funds of the order of £15 million by the early 1990s to keep development work on track.

INTERNATIONAL COLLABORATION ON HYPERSONIC PROJECTS

In view of the heavy French and West German financial commitment to the Ariane-5, Hermes and Columbus programmes and the British government's lack of conviction over the case for Hotol, it was clear that Western Europe would be extremely hard-pressed to pay for one hypersonic plane project by the turn of the century, let alone two.

There was no prospect of the Hotol and Saenger projects being built side by side – if anything was to be built at all, agreement would have to be reached on which basic concept offered the best prospect of success. The fundamental choice between a one-stage and two-stage launch vehicle would need to be made, and that choice would almost certainly be based not simply on technological or commercial grounds, but would also reflect the political backing behind the rival projects.

The Hotol project was attractive in its single-stage simplicity, but there were fears in the industry that the attempt to build an engine running on both air and liquid oxygen could prove considerably more difficult and expensive than anticipated. The Saenger project, by separating out the propulsion problems of the atmospheric ascent phase and the orbital phase and developing radically different engines for each of the two vehicles, appeared to be taking a less risky approach. At the same time, because of the similarities between the Hermes shuttle and the Horus upper stage of Saenger, a considerable part of the research and development work for Horus was already *de facto* being paid for by ESA countries committed to the Hermes programme.

Given the huge costs of a hypersonic plane programme, estimated at anywhere between $10 and $20 billion depending on the design chosen, it was also possible that the Western world as a whole would be incapable of funding more than one such project, and that the American, Japanese and European industries would eventually find it in their interests to club together to build an international space plane, much as they had already agreed to build an international space station.

Elements of NASA's National Aerospace Plane project, Hotol, Saenger and Japanese projects could then all eventually be combined to produce a blueprint for a programme that would be both affordable and contain the best elements of each design. Tremendous hurdles would have to be crossed, however, if such a solution were to be reached.

The experience with the space station negotiations did not augur well for a transcontinental approach. Rival aerospace groups from different countries would compete fiercely for plum contracts on the plane, arguments could arise over whether it should be designed primarily with civilian or military users in mind and nationalistic prejudices could dictate a division of labour that did not in the end produce the best product. But ultimately the sheer scale of the project, and the cost running into tens of billions of dollars, could dictate a co-operative effort.

Even on the most optimistic assumption, a hypersonic space plane was unlikely to be operational before the year 2000. In the intervening period the drive to make conventional launch systems, whether expendable rockets or semi-reusable shuttle systems, was certain to step up as satellite owners and others interested in launching payloads had an increasing range of options from which to choose. And as a result they were beginning to voice their concerns more loudly over the launch conditions being offered, the prices they had to pay, and another factor that was looming increasingly large in financial calculations – the kind of insurance cover that they could buy to cover themselves against launch failures.

LAUNCH COSTS AND SATELLITE INSURANCE

In the early days of the space programme satellite launches were frequently offered free or at low cost. When rockets were being developed and perfected it cost little more to put a satellite on board, adding an extra dimension to the test flights.

As the 1970s progressed and satellite owners began to be faced with increasingly high prices for launches, they started to look around for insurance cover. Underwriters, inexperienced in the business and impressed by the growing reliability of American rockets, whose launch success rate was generally well above 90 per cent, started by offering unrealistically low premiums of between 6 and 10 per cent of the launch cost. In the period up to 1982 underwriters worldwide collected premiums totalling around $100 million, but had to pay out twice that amount in insurance claims relating to the loss of seven satellites. Their fingers badly burned, they pushed rates up to between 12 and 15 per cent in 1984 and 20 per cent in 1985, both disastrous years for the space industry during which claims for lost satellites totalled more than $620 million. Cumulative losses by the insurance industry up to the end of 1985 were in excess of $500 million dollars.[42]

With satellites frequently costing around $80 million each to build and another $50 to $60 million to launch, operators were understandably unwilling, and frequently unable, to contemplate paying a further $20 to $30 million per satellite in insurance fees. As a result the authorities responsible for launch systems started to offer their own insurance schemes, sometimes taking the form of cash guarantees and sometimes free or cut-rate repeat launches of back-up satellites in the event of failure of the primary launch.

NASA, prompted by a series of satellite launch failures and the embarrassment of having three satellites carried into space by the shuttle subsequently stranded in useless low Earth orbit, was offering a scheme by late 1985 which guaranteed a free repeat launch if the shuttle's first or second ascent stages failed and a half-price repeat launch if problems arose either with the module used to place the satellite in transfer orbit or if the satellite failed within its first three months in orbit.

Arianespace's response was to set up its own insurance company and, calculating premiums on the basis of one launch failure every 15 flights, offer coverage totalling $73 million in cash or a free reflight. Premiums were set at 11 per cent of launch cost for those opting for a reflight, or 13.2 per cent for those wanting cash coverage. At the time the estimated failure rate of one flight in 15 was based not on the performance of the Ariane rocket to date, but instead on a projected future improvement in that performance, assuming a similar increase in reliability to that seen on the older US rockets. In fact three out of the first 14 Ariane flights after its initial test phase ended in failure, or two out of the first ten flights carried out following the creation of Arianespace as a commercial company, giving the rocket an early reliability rate of barely 80 per cent. Arianespace's faith in an improved launch reliability seemed to have been justified by the start of 1990, when its success rate had risen to about 90 per cent.

The Challenger accident in January 1986, the failures of US Delta and Titan rockets a couple of months later and the Ariane V18 accident in May that year not only crippled Western launch programmes for more than a year but knocked the bottom out of the satellite insurance market.

Operators seeking quotations for launches in the late 1980s and early 1990s were being asked to pay 25 per cent or more of the launch cost, a realistic price based on risk-assessment from the insurer's point of view but totally unacceptable for commercial satellite operators. As a result private US launch companies started offering their own free re-flight schemes, while some satellite owners initiated a form of self-insurance, buying more satellites than they needed.

Andrea Caruso, Eutelsat director-general until his resignation in 1989, said the organisation assumed one failure in four launches and built this into its satellite orders.

> Therefore if we need three, we buy four. If we need four, we buy five ... the only kind of insurance that we accept is that of the launcher itself, because it is offered to us at reasonable conditions. But insurance as far as the satellite and its performance are concerned is still prohibitive ... Moreover, I don't need money, I need satellites ... Continuity is the password in telecommunications. The insurance does not protect you against the interruption of services.[43]

It was not just insurance which provoked criticism from satellite operators, however, but also the pricing policies of the launch companies. Three major complaints topped the list. Firstly, satellite owners were unhappy with the 'best effort' clause in launch contracts, which meant that launch companies simply guaranteed to do their best to launch a satellite but did not guarantee placing anything in orbit. Secondly, they felt that the system of paying advance instalments of launch fees was unfair. And thirdly, they disputed the system under which they had to pay penalties for delaying a launch, while receiving no compensation from launch companies if they suffered launch delays due to problems with the rockets.

> I cannot accept anymore this principle of paying $50 to $60 million for satellite launches for 'best efforts'. I am not happy at all with the terms and conditions that are generally applied by the launch service suppliers

Caruso complained.[44]

And Allan McCaskill, manager of Intelsat's launch vehicle programmes, echoed his sentiments, saying that in his experience in over 20 years of dealing with launch companies, 'the only thing certain was that we paid our money'.[45]

The response of the rocket companies was that, despite improvements in launcher success rates, it was not possible to give 100 per cent guarantees for a service that was still not 100 per cent reliable, and that advance payments helped keep down launch costs, so that any major change in the system would probably lead to higher prices for satellite operators.

Whereas in the early 1980s there were only two major satellite launch systems available to Western customers, Ariane and the US shuttle, by the end of the decade the choice was far greater. The change in US policy following the Challenger accident meant that three major aerospace contractors, McDonnell Douglas, General Dynamics and Martin Marietta, had entered the arena with commercial versions respectively of their already well-proven Delta, Atlas Centaur and Titan rockets. Not only that, but the Soviet and Chinese authorities were also beginning to offer cut-price rival services of their own.

SOVIET AND CHINESE LAUNCH COMPETITION

Both China and the USSR were offering Western clients launches for around $30 million, little more than half the cost of launching on the major Western rockets. The Soviet Union was offering launches on its Proton rocket, the workhorse of its space programme which had a long and solid track record of reliability, while the Chinese were offering launches on their Long March rocket. The US attitude to the two countries was of key importance, since many satellites were either designed and built under the prime contractorship of a US company, or had sensitive US electronic components, which meant that US government permission had to be given before they could be transported to a Soviet or Chinese launchpad. US opposition to Soviet launches of Western satellites was resolute, severely limiting Moscow's potential market. The situation was different with China, however, at least prior to the student protests in Tiananmen Square in May 1989 and the brutal crackdown which followed in June. In December 1988 the US and China initialled an agreement under which China could launch up to nine US commercial satellites in the six years to 1994, amounting to an average of 1.5 satellites per year. The move, while granting Peking permission for such flights, also effectively put an upper limit on the competition that Western launch operators might expect from that quarter. And China agreed not to over-subsidise its launches, bringing its prices more into line with those of its Western rivals.

SMALL SATELLITE LAUNCHES

Aside from competition among the large launch companies to lure operators onto their rockets, space engineers began identifying a gap in the market for small satellites in the 500 kilo to one tonne range, and micro-satellites. These could be launched as subsidiary payloads to bigger satellites on large rockets, but their owners had few rights to determine launch dates, and their needs were totally subordinated to those of the larger satellite operators who were essentially paying for the launch.

Since the late 1950s a small rocket called the Scout, developed jointly by LTV of the US and Snia BPD of Italy, had been conducting launches of scientific payloads from the Italian San Marco platform off the coast of Kenya as well as two launch sites in the US, Vandenberg Air Force Base, California and Wallops Island off the east coast. Costing around $12 million per launch, it had accomplished more than 100 flights by the late 1980s with a 95 per cent reliability record, carrying payloads to study everything from solar physics and astronomy to auroral phenomena and meteorology. While the first Scout could place a 60 kilo payload into orbit at an altitude of 555 kilometres, by the late 1980s it could carry three times that mass. An improved version, Scout II, putting more than 500 kilos into orbit, was under study and due for a first launch in 1993.

A rival system to the Scout, LittLEO, was also under development in Britain. Conceived at first as a vehicle to put payloads of 300 kilos into a 300-kilometre circular orbit, the design gradually changed to accommodate payloads of first 500 kilos and then a tonne. Market surveys conducted by the LittLEO promoters suggested that there was demand from national and international agencies and industry for a vehicle capable of launching small payloads into low Earth orbit at a price of around $10 000 per kilo, with a total European market during the 1990s of between 50 and 100 small satellite launches.

While Scout was flying, LittLEO (the name signifying a little rocket flying into Low Earth Orbit) was still on the drawing board at the end of the 1980s. But plans were moving ahead for an inaugural launch from a range at Andoya, Norway in 1991, using initially US-made solid fuel boosters. LittLEO executives said that the price of $10 000 per kilo indicated by market surveys could be difficult to offer from a Scandinavian launch site, but that it could be attainable with an equatorial base where less fuel would be needed to achieve orbit.

Meanwhile, Arianespace was offering limited space aboard its rockets to people wanting to launch a range of micro-satellites, weighing typically less than 50 kilos, small satellites in the 200 kilo plus range, technology experiments generally weighing somewhere between the two, and micro-gravity experiments. Arianespace's policy on such launches, where the small customers were squeezed into spare space alongisde a larger satellite, was to charge simply to cover costs, but to accord small passengers no rights to interfere with the main launch mission.

The launch market, having become essentially a two-horse race in the early 1980s between Ariane and the shuttle, by the end of the decade had been totally transformed. Competition was much more evident, with no less than six separate companies or agencies vying for Western satellite launches, and more were springing up. Arianespace, concerned that launch prices were especially open to hidden subsidy, was urging the establishment of a global launch pricing policy, a plea which appeared to be falling on deaf ears. Competition on price suited both those who had payloads to launch and the newcomers to the commercial launch market, who had clients to win and a niche to establish. It was no longer adequate simply to offer a service which put a payload into space – the price and conditions of launch were becoming a key concern in the new, more competitive environment.

15 The International Scene and the Way Forward

European collaborative space efforts, as well as the individual national space programmes, needed to be judged not only on their own merits but also against what was happening in other countries. From the early 1960s Europe had been running a distant third in the space race, well behind the US and Soviet Union, with many Europeans totally unaware even of the existence of a European space programme.

France had emerged in the 1970s as the undoubted political leader of the European space community, and by the late 1980s had an annual budget of more than $800 million, well above West German space funding and five times the amount that Britain was spending. While there were disagreements within ESA over priorities, Britain was alone within the organisation in questioning the whole thrust of its programme and casting doubt on the value of an expanding state-funded space programme. The British attitude was also unusual when viewed in a more international context.

THE UNITED STATES – THE RIDE REPORT

In the US NASA had passed through a critical period following the Challenger accident, when many of its past achievements were buried in a wave of recrimination over management failure and lack of accountability. The Rogers Commission, appointed by President Reagan to investigate the shuttle disaster, while criticising past NASA actions did point the way to reforms which could put the space programme back on track. And in August 1987 NASA published the Ride report, setting out a range of suggested projects to restore a sense of purpose to US space policy, ranging from Earth observation programmes to the exploration of the deepest reaches of the solar system, setting up a permanent manned base on the moon and sending astronauts to Mars. Important as the details of the report were, its significance lay also in the positive, forward-looking approach it took after the traumatic introspection of the post-Challenger period. 'The US civilian space programme is now at a crossroads', it warned, calling for bold initiatives staggered over a sufficiently long timescale to make their multi-billion dollar budgets begin to look affordable.

Setting out in enthusiastic terms plans for a Mars landing in the first decade of the next century, paving the way for a permanent manned presence on the planet, the report sought to justify the programme as much in terms of lifting national morale as reaping scientific results.

191

A successful Mars initiative would recapture the high ground of world space leadership and would provide an exciting focus for creativity, motivation and pride of the American people. The challenge is compelling, and it is enormous.[1]

So was the cost, estimated by the press at anything up to $75 billion.

NASA had no shortage of critics in the Administration and Congress, but even the Rogers Commission, which pulled no punches in attacking NASA safety standards and management procedures, recommended that it should continue to receive full government support as 'a symbol of national pride and technological leadership'.[2]

The shuttle's return to space on 29 September 1988 after nearly three years in the hangar demonstrated the depth of popular enthusiasm for space spectaculars in the US, as well as the extent to which emotional scars from the Challenger explosion remained unhealed. Hundreds of thousands of people lined the beaches and roads near the launch site, millions watched the launch live on television. 'The magic is back!' trumpeted *Time* magazine, while the electronic board in New York's Times Square triumphantly flashed the message 'America returns to space' to the crowd below.[3]

The resurgence of enthusiasm for space in the US, although restrained by concern in Congress and elsewhere over the cost of the programmes, was reminiscent of the mood during the Apollo moon programme. President Bush's decision, following his inauguration in January 1989, to give Vice-President Dan Quayle personal responsibility for US space policy, while not guaranteeing that major projects would be pushed forward, at least helped to underscore the importance that the Administration felt the issue deserved to be given publicly.

THE SOVIET UNION – THE GORBACHEV ERA

While attitudes to space in the US were evolving from the indifference prevalent throughout much of the previous two decades, the situation in the Soviet Union was also changing. One innovation during the Gorbachev era was the live televising of Soviet launches, unheard of in earlier times when the entire programme was shrouded in secrecy and missions sometimes were only reported after they had ended. The new openness reflected not only a changed political climate in Moscow but also the great confidence that Soviet engineers had in their launch systems. Technologically the Soviet rockets, capsules and even space stations were not particularly advanced, making use of well-tested designs which guaranteed uninterrupted access to space for both cosmonauts and unmanned spacecraft. The Soviet Union was playing tortoise to the American hare, and as in the fable the tortoise was pulling ahead. The Ride report identified two areas in which the Soviet Union was clearly leading NASA – the exploration of the Martian surface by robots, and the space station programme which appeared likely to give Soviet cosmonauts more than a decade's experience of long-duration space flight before the US-led international space station could be in orbit and operational.

True, America had won the race to the moon, but there had been no follow-up lunar programme in which it could consolidate its expertise in that area. US technology was generally superior to that of the Soviet Union, but its potential was not being fully exploited in space. And in the one area where the US had indisputable leadership, the shuttle, the Soviet Union had caught up fast. The November 1988 unmanned first flight of the Buran had demonstrated just how determined Moscow was to stay competitive in space.

THE JAPANESE CHALLENGE

Other nations apart from the Europeans were beginning to challenge Soviet and American dominance of space in a variety of areas ranging from launchers to satellite systems. Both Japan and China had active space programmes, China's having reached a greater level of maturity but Japan's showing more technological potential. By the late 1980s the Japanese government was spending over 110 billion yen ($500 million) a year on space, largely through the National Space Development Agency of Japan (NASDA) which had been set up in 1969. Its N-I rocket, which was essentially an adaptation of the McDonnell Douglas Thor Delta, made seven flights between 1975 and 1982, carrying very small payloads of up to 130 kilos into geostationary orbit. An improved version called the N-II, a rocket still based on US technology with three stages and nine boosters, flew eight times between 1981 and 1987, putting payloads of up to 350 kilos into orbit.

However, the breakthrough in Japanese rocket technology came on 27 August 1987 when the first H-I rocket was successfully launched from the Tanegashima space centre on the southernmost tip of Japan. Carrying a 550-kilo telecommunications satellite, the rocket was 84 per cent manufactured in Japan, with domestic rocket scientists notably responsible both for its cryogenic second stage and solid fuel third stage, as well as its inertial guidance system. Its first stage motor was identical to the N-II first stage. Like the N-I, it had three stages and nine boosters, and was scheduled to make eight flights up to 1991. Japanese launch frequency was partly dictated by an agreement with the United States limiting the development of potentially offensive military technology, and partly by the requirements of the fishing industry near the launch site. As a result the space authorities were restricted in practice to two launches each year.

The H-I represented Japan's transition from launching an essentially American rocket to a home-produced one. The next big step, scheduled for 1991 or 1992, would be the inaugural flight of the H-II, a rocket similar in concept to Europe's Ariane-5 project, although much smaller. It would be a two-stage vehicle, using a large liquid oxygen and liquid hydrogen motor for the first stage and a smaller cryogenic system for the second. This rocket would be able to put a single two-tonne satellite, or multiple payloads totalling two tonnes, into geostationary orbit, four times the capacity of the H-I. It would take satellites up to 4.6 metres in diameter, making it compatible with the US shuttle, and would also be able to launch deep space probes.

Aside from the rocket programme, Japan had also been steadily working on a range of satellite projects. The seven N-I flights put three technology orbiters, two Earth observation satellites and two telecommunications satellites into orbit while the N-II launched a further three Earth observation vehicles, four communications and broadcast satellites and another technology orbiter. The Earth observation programme included the GMS family of meteorological satellites, the first launched on an American Delta in 1977 but subsequent ones launched from Tanegashima, making a big contribution to weather forecasting not only around Japan but over a wide swathe of the Far East and the Pacific.

Japan's MOS-1 observation satellite, launched in February 1987, was designed to collect data on water vapour in the atmosphere, ocean surface temperature, currents and ice movements as well as information on mineral deposits and crops. Japan also processed data for its own use from the US Landsat remote sensing programme.

Japanese plans for future programmes included the construction of a shuttle, which would be modelled on the US version but would be initially unmanned and considerably smaller. It would be launched on top of the H-II, rather as the European Hermes would be designed to ride on top of Ariane-5. A more sophisticated version, requiring a more powerful H-II with multiple solid rocket boosters, would in theory fly later in the 1990s, paving the way for the development of a horizontal take-off space plane in the early part of the next century.

Trial tests of a mini-version of the shuttle got off to an inauspicious start in September 1988, however. A two-metre-long model of the future spacecraft, sitting on a small rocket, was due to float up to an altitude of 20 kilometres suspended from a helium balloon before blasting off and then coasting back down through the atmosphere to test its aerodynamic performance. The model, which cost $2 million to construct, never made it however, the helium balloon puncturing at an altitude of 18 kilometres and dropping the shuttle unceremoniously into the Pacific. It was a frustration rather than a major setback, but unusual in that the Japanese space programme, by following cautiously where other nations had made the running, had managed to steer remarkably clear of mishaps.

Apart from rocket, shuttle and satellite programmes, Japan was also developing its interest in science in orbit. It planned to launch a free-flying unmanned laboratory in early 1993, designed like Europe's Eureca project to stay in low Earth orbit for about six months. It would be launched by an H-II rocket and retrieved by a US shuttle for analysis back on Earth.

Japanese industry was working on various concepts for space platforms in the late 1990s combining both telecommunications and scientific functions. And Japan was also contributing to the US-led international space station, building an experimental module for attachment like ESA's module to the main space station structure.

CHINA – A POLICY CONUNDRUM FOR THE US

While Japan was forging ahead with a space programme based on a mixture of US and homespun technology, China had been working with little fanfare for two decades to build up its own space expertise. It launched its first satellite in 1970 and over the following 20 years launched more than 20 others, using its family of Long March rockets. Only a passing interest was taken in the West in China's rocket programme, which was clearly well behind that of the US and Soviet Union and relied on fairly classic rocket technology. When China started offering its launch services to Western satellite operators in the mid-1980s, however, both the space industry and governments began to take notice. The simultaneous grounding of both the shuttle and Ariane put the spotlight briefly on China, and it won a contract to launch a Swedish scientific satellite, Fleya, in 1991 as well as expressions of interest from other satellite operators such as Western Union Corp and Pan Am Pacific. US Defense Secretary Caspar Weinberger visited the Xichang launch site in 1986, and there followed a period of protracted negotiations between the US and Chinese governments over the circumstances under which Washington would permit Chinese rockets to launch satellites containing US technology. The Reagan administration had been concerned both to prevent a flow of technology to China through satellite transit operations, and to protect the nascent private US launcher industry, which the government was encouraging to grow in the wake of the Challenger accident and the decision to switch commercial satellite payloads off the shuttle.

In December 1988 China and the US signed an accord under which China agreed to restrict its launches of US commercial satellites to a total of nine in the period 1989/94, equal to a rate of 1.5 satellites per year. China also agreed to charge launch prices comparable to those being asked by Western companies, rather than undercutting the competition to attract foreign hard currency. It remained to be seen whether China would actually manage to sell its nine launch slots, given the expanding launch capacity in the US with Delta, Titan and Atlas Centaur rockets all competing for customers, in addition to Ariane. The dramatic change in the political climate in China following the crushing of the student demonstrations in early June 1989 also cast serious doubt over the country's ability to attract satellite launch customers.

China was offering launches on its Long March 2 rocket, which could put up to 1.25 tonnes into geostationary orbit or 2.4 tonnes into low Earth orbit, or the more powerful Long March 3, which could put 1.4 tonnes into geostationary orbit. Chinese rockets planned for the early 1990s would be able to put as much as nine tonnes into low Earth orbit or four tonnes into geostationary transfer orbit.

Another area which China was interested in exploiting commercially was that of launching experimental packages into space for short periods and then recovering them. French electronics group Matra, frustrated at the grounding of the shuttle, became the first Western company to use this facility in August 1987, flying two micro-gravity experiments in space for five days. The Chinese authorities subsequently said that more than ten other Western companies had expressed interest in flying experiments aboard Long March rockets.

SMALLER SPACE POWERS

Behind the major space nations a host of minor ones were jostling for position. Australia, which already had a telecommunications satellite system in place, was not attempting to develop its own rockets but was interested in taking advantage of its geographical position to build an international space Port on the Cape York peninsula in the far north of Queensland, barely 10 degrees from the equator and ideally situated for satellite launches. Both the Soviet Union and China, hampered by their positions far from the equator, expressed early interest in using the launch facility, if built.

India joined the élite club of space nations when it successfully launched two small satellites, each weighing less than 50 kilos, into orbit in 1980 and 1983. It subsequently began work on developing a five-stage rocket to place a 150-kilo payload into orbit, in parallel with its military interest in developing long-range missiles. But the first two launch attempts of the new rocket model, one in early 1987 and the second in mid-1988, both ended in failure when the rockets plunged into the Bay of Bengal soon after lift-off.

Israel shocked the Arab world in September 1988 with its first satellite launch, a small technological satellite weighing barely 150 kilos, Ofek-1, aboard a rocket called Shavit. The satellite itself was experimental, but was viewed in neighbouring Arab countries as the forerunner of military spy satellites which they feared would give Israel even greater power in the region than it already wielded. Israel's space agency director Yuval Neeman conceded the military potential of the project when he commented following the launch that

> there is defence potential in all of this action, but that is for the defence establishment to decide. Now the importance of the satellite is technological.[4]

Brazil too was working on its own launcher project, although it was unclear when this might become operational, while several Asian, Arab and Latin American countries had formed groups to acquire telecommunications satellite networks from US or European suppliers.

BRITISH LACK OF DIRECTION

What was obvious, even from a fairly superficial survey of global programmes, was that interest in exploiting space was growing rapidly in both the developed and developing world among those countries which already possessed technology of a sufficiently high standard to pursue their own projects and those which were anxious to acquire it. This was what made the attitude of the British government so unusual. As the 1980s drew to a close Britain was the only country out of the five major nuclear powers – the other four being the US, Soviet Union, China and France – to be spending virtually no government money on launch systems.

Among the Group of Five leading industrial nations in terms of economic power – the US, Japan, West Germany, France and Britain – the Thatcher government was similarly isolated in its belief that rockets were simply not worth spending money on. Of course, being in a minority of one did not necessarily mean that Britain was wrong. But it certainly posed questions about the wisdom of the government's view that state spending on space should be determined by the same cost/benefit criteria applied to other areas of national life, despite the newness of the technology, the large investments involved and the long time lag between a project's experimental and commercial phase. British Trade and Industry Secretary Lord Young laid out the government's market-oriented space policy in stark terms in a House of Lords debate in March 1988:

> The single test that we should surely apply in government is not that of inputs (how much things cost) but of outputs (what we get). That is the one we must bear in mind ... we reserve the right to apply the same economic tests and the same spirit of looking for outcomes to space as we have in the rest of the economy. In this field space is on land as well as up in the air

he said.[5]

The House of Lords select sub-committee on space, under the chairmanship of Lord Shackleton, while refusing to be carried away by dreams of space projects beyond the national budget and even taking the line that it was not necessary for Europe to develop its own manned space capability independent of the Americans, nevertheless urged a more visionary approach to the question of space funding than that shown by the British government. It urged the formulation of a much clearer space policy, and stated bluntly that the bulk of increased spending would have to come from the public sector rather than industry.

> Man has fought for control of the land, the sea and the air, and he is now tackling the challenge of space. The United Kingdom has to accept this challenge, as the other major industrialised countries are doing. The reasons are political, cultural, military, economic, commercial and scientific; these arguments stand together

the committee wrote in its report.

> The space race will not be won simply by the first nation to reach Mars or to put a family of four into orbit. There is no finishing line. But there are plenty of prizes, and the major technological powers in the world are making up their teams for the race. The United Kingdom has the chance to join in. Unless we do so now, we shall never be in the running.[6]

PRIVATE INITIATIVES – CHANGING ATTITUDES TO SPACE

While the argument raged in Britain over whether it was necessary at all to increase its modest space spending, elsewhere the pace was being forced not only by governments and major aerospace firms but also by a growing number of private individuals and small companies anxious to cash in on the commercial possibilities. Many of these had been encouraged by President Reagan's invitation to private US industry to take over the role of launching civilian payloads from the shuttle, and the new encouragement of private enterprise involvement had a knock-on effect at many levels of the industry.

One of the more macabre proposals to come out of the US was a project evolved by a Florida group of space engineers and undertakers to launch a satellite containing the remains of several thousand dead people, at a cost of around $3000 per person, an orbiting mausoleum designed to be seen on clear nights as it passed overhead, glinting in the sky to the memory of those aboard.

On an even grander scale, the Eiffel Tower company in Paris launched a competition to find a space project that could be put into orbit to mark the centenary of the Eiffel Tower's construction in 1889, a project that was supposed to embody the spirit of the original venture. A jury picked a short-list of three proposals in October 1986 from 99 submitted, and the final winner was a French-designed 'ring of light' comprised of an inflatable plastic ring 24 kilometres in circumference with 100 spheres strung along its length at 240-metre intervals, designed to reflect the sunlight during night-time on Earth. Its circumference was the same as that of the historic centre of Paris, and to the observer on the ground it would appear about the same size in circumference as the moon in the sky. Arianespace was approached about the possibility of a free launch for the inflatable package, but strong opposition, primarily from astronomers, eventually forced the project to be abandoned. Astronomers feared that highly-sensitive telescopes, designed to probe deep into the universe and register tiny quantities of light emanating from distant galaxies and solar systems, would be severely damaged if the bright orbiting necklace of light crossed their field of view.

The project, although never completed, highlighted the fact that space, for a long time the preserve of scientists, governments and major aerospace firms, was becoming accessible to a new group of users. With falling launch costs, commercial initiatives and even space advertising began to come potentially within corporate budgets. And there was a growing pool of engineers with space experience eager to escape the sometimes monolithic space organisations and set up on their own. Former NASA astronaut Deke Slayton, head of the Houston-based Space Services Inc, launched a 37-foot rocket 195 miles into space from a pad on Matagorda Island, Texas, in September 1982. In Europe the LittLEO team were hoping to find a niche in the private launch market for small satellites launched from Norway. And a Florida space engineer, Bob Davis, set up a company called EPrime Aerospace Corp to launch 14-foot rockets out over the Atlantic, providing around one minute of weightless flight for microgravity experiments in the first ever purely private space venture.

Such initiatives were bound to highlight the lack of international law governing space, raising questions such as whether the scientific community should automatically take precedence over people or corporations wanting to launch 'space art' or advertising material, and whether the interests of the big powers could peacefully co-exist in space alongside those of the private sector being encouraged to climb aboard.

Ultimately the question 'to whom does space belong?' would have to be addressed. The search for an answer could keep droves of lawyers, scientists, artists, politicians and philosophers arguing happily for decades. In the meantime the answer was likely to remain 'to those who are prepared to pay to get there'. In that respect the major nations of Europe, with the possible exception of Britain, seemed to have voted with their feet, and their pocketbooks, and signed up for the ride.

Appendices

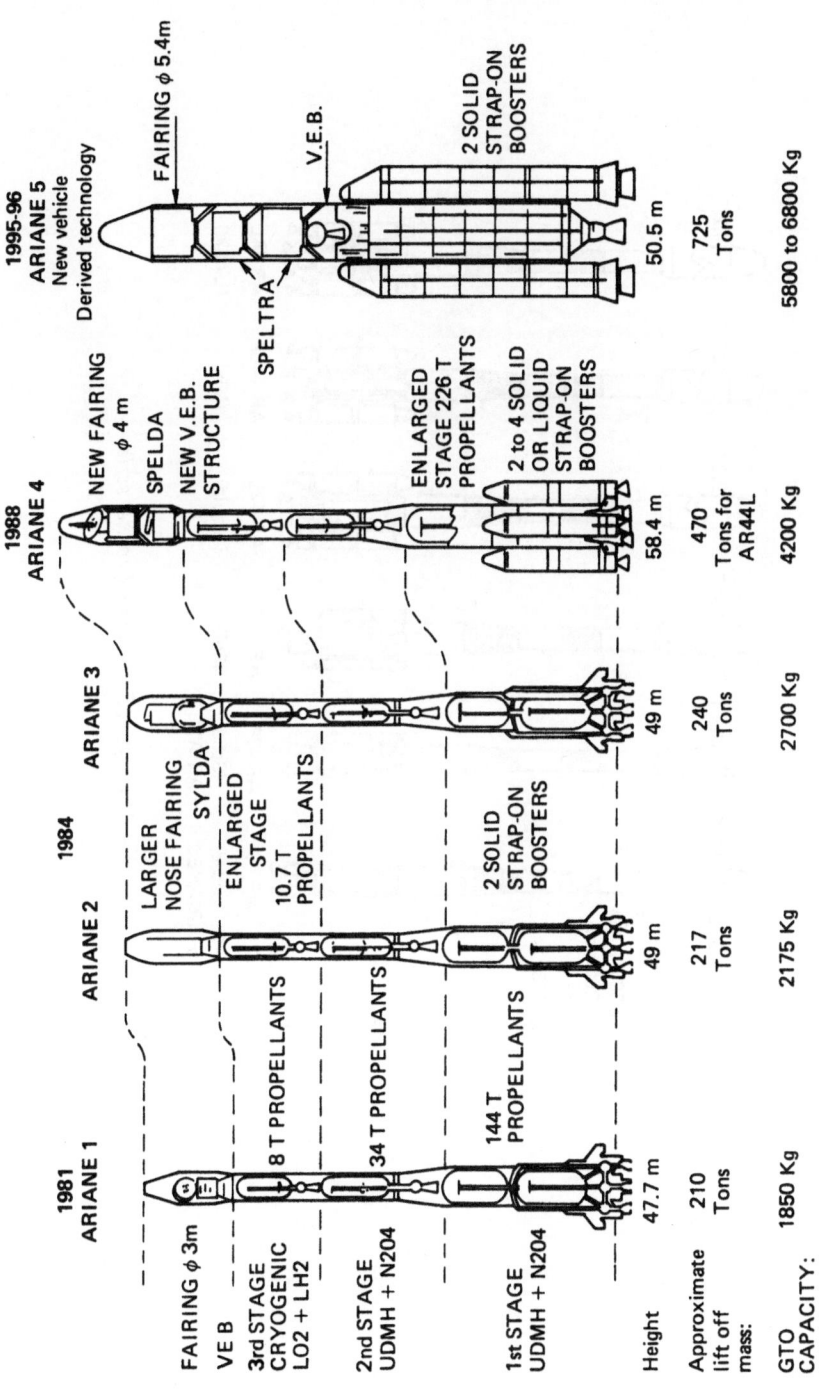

APPENDIX 1 *Ariane rocket family*

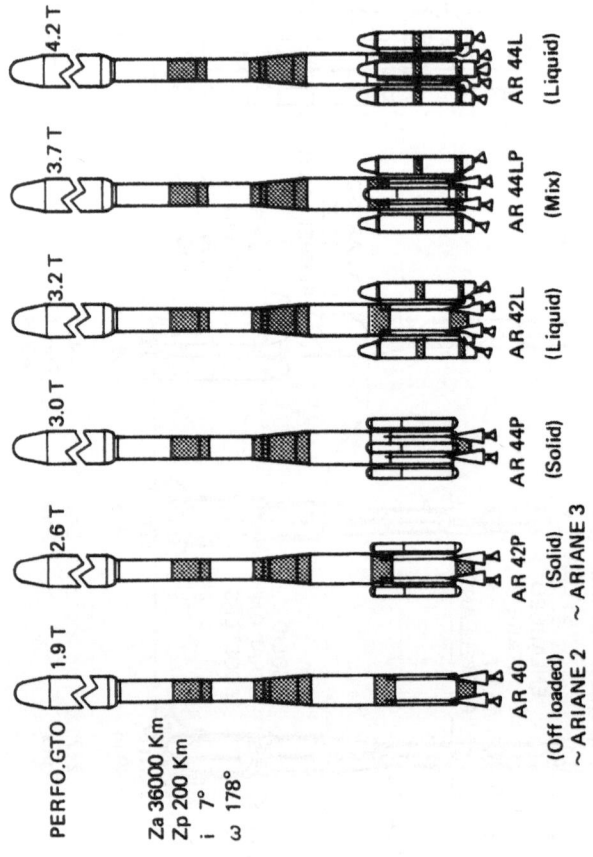

APPENDIX 2 *Ariane 4 rocket series*

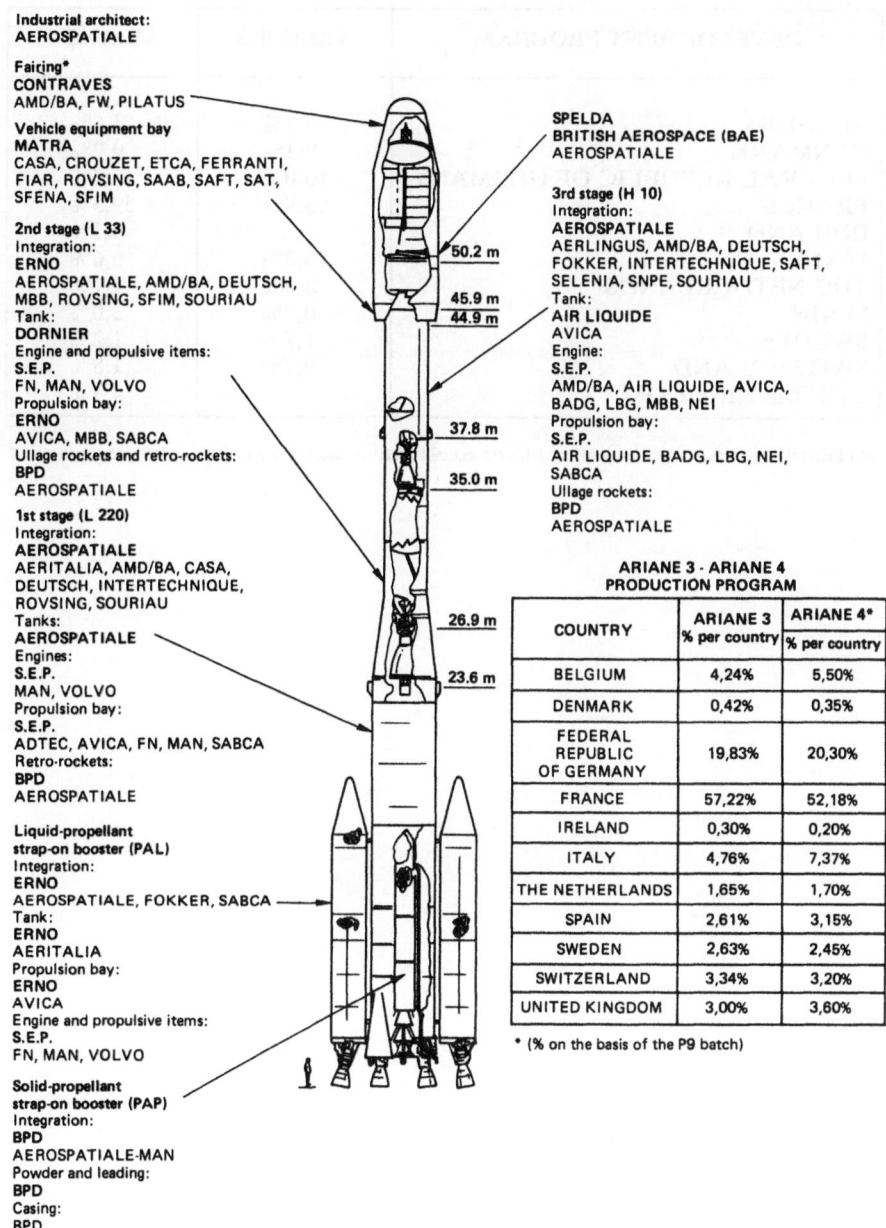

Industrial architect:
AEROSPATIALE

Fairing*
CONTRAVES
AMD/BA, FW, PILATUS

Vehicle equipment bay
MATRA
CASA, CROUZET, ETCA, FERRANTI,
FIAR, ROVSING, SAAB, SAFT, SAT,
SFENA, SFIM

2nd stage (L 33)
Integration:
ERNO
AEROSPATIALE, AMD/BA, DEUTSCH,
MBB, ROVSING, SFIM, SOURIAU
Tank:
DORNIER
Engine and propulsive items:
S.E.P.
FN, MAN, VOLVO
Propulsion bay:
ERNO
AVICA, MBB, SABCA
Ullage rockets and retro-rockets:
BPD
AEROSPATIALE

1st stage (L 220)
Integration:
AEROSPATIALE
AERITALIA, AMD/BA, CASA,
DEUTSCH, INTERTECHNIQUE,
ROVSING, SOURIAU
Tanks:
AEROSPATIALE
Engines:
S.E.P.
MAN, VOLVO
Propulsion bay:
S.E.P.
ADTEC, AVICA, FN, MAN, SABCA
Retro-rockets:
BPD
AEROSPATIALE

Liquid-propellant
strap-on booster (PAL)
Integration:
ERNO
AEROSPATIALE, FOKKER, SABCA
Tank:
ERNO
AERITALIA
Propulsion bay:
ERNO
AVICA
Engine and propulsive items:
S.E.P.
FN, MAN, VOLVO

Solid-propellant
strap-on booster (PAP)
Integration:
BPD
AEROSPATIALE-MAN
Powder and leading:
BPD
Casing:
BPD
FBM

SPELDA
BRITISH AEROSPACE (BAE)
AEROSPATIALE

3rd stage (H 10)
Integration:
AEROSPATIALE
AERLINGUS, AMD/BA, DEUTSCH,
FOKKER, INTERTECHNIQUE, SAFT,
SELENIA, SNPE, SOURIAU
Tank:
AIR LIQUIDE
AVICA
Engine:
S.E.P.
AMD/BA, AIR LIQUIDE, AVICA,
BADG, LBG, MBB, NEI
Propulsion bay:
S.E.P.
AIR LIQUIDE, BADG, LBG, NEI,
SABCA
Ullage rockets:
BPD
AEROSPATIALE

50.2 m
45.9 m
44.9 m
37.8 m
35.0 m
26.9 m
23.6 m

ARIANE 3 - ARIANE 4
PRODUCTION PROGRAM

COUNTRY	ARIANE 3 % per country	ARIANE 4* % per country
BELGIUM	4,24%	5,50%
DENMARK	0,42%	0,35%
FEDERAL REPUBLIC OF GERMANY	19,83%	20,30%
FRANCE	57,22%	52,18%
IRELAND	0,30%	0,20%
ITALY	4,76%	7,37%
THE NETHERLANDS	1,65%	1,70%
SPAIN	2,61%	3,15%
SWEDEN	2,63%	2,45%
SWITZERLAND	3,34%	3,20%
UNITED KINGDOM	3,00%	3,60%

* (% on the basis of the P9 batch)

*For each element, the prime contractor's name comes first (in bold type) followed by the list of sub-contractor's

APPENDIX 3   Ariane 4, showing industrial contributors

DEVELOPMENT PROGRAM	ARIANE 3	ARIANE 4
BELGIUM	0,8%	4,6%
DENMARK	0,1%	0,2%
FEDERAL REPUBLIC OF GERMANY	16,0%	18,2%
FRANCE	55,9%	59,3%
IRELAND		0,1%
ITALY	14,3%	6,6%
THE NETHERLANDS	0,8%	1,1%
SPAIN	0,7%	2,0%
SWEDEN	1,7%	1,2%
SWITZERLAND	9,7%	1,8%
UNITED KINGDOM		4,9%

APPENDIX 4 *National Contributions to Ariane 3 and 4 development programmes*

COMPANY	COUNTRY	ARIANE 2-3	ARIANE 4
AEROSPATIALE	France	Industrial architect of the Ariane systems First and third stages Pyrotechnical separation systems SYLDA (Dual Launch System)	
AIR LIQUIDE	France	Third stage fuel tank	
BRITISH AEROSPACE	United Kingdom		SPELDA (Payload bearing structure for multiple launches)
CONTRAVES	Switzerland	Fairing	
MATRA	France	Vehicle equipment bay	
MBB-ERNO	Germany	Second stage	Second stage Liquid propellant boosters (LPB)
SEP	France	Prime contractor for the propulsion systems First, second and third stage propulsion systems	First, second and third stage propulsion systems Liquid-propellant booster propulsion systems
SNIA BPD	Italy	Solid-propellant boosters (SPB) Ullage and retro-rockets	

APPENDIX 5 *Ariane industrial programme share out*

DATE	FLIGHT	LAUNCH VEHICLE	SATELLITES	CUSTOMER
1979 Dec. 24	L01	AR1	CAT (technological)	
1980 May 23	L02*	AR1	CAT	
			FIREWHEEL	Max Plank Institute (FRG)
			AMSAT	AMSAT
1981 June 19	L03	AR1	CAT	
			APPLE	ISRO (India)
			METEOSAT	ESA
1982 Oct. 20	L04	AR1	MARECS A	ESA for INMARSAT
1982 Sept. 10	L5*	AR1	MARECS B	ESA for INMARSAT
			SIRIO-2	CNR (Italy)
1983 June 16	L6	AR1	ECS1	ESA for EUTELSAT
			OSCAR 10	AMSAT
1984 Oct. 19	L7	AR1	INTELSAT V (F7)	INTELSAT
1984 Mar. 5	L8	AR1	INTELSAT V(F8)	INTELSAT
1984 May 22	V9	AR1	SPACENET 1	GTE SPACENET (USA)
1984 Aug. 4	V10	AR3	ECS2	ESA for EUTELSAT
Nov. 1	V11	AR3	TELECOM 1A	CNES for France TELECOM (Fr.)
			SPACENET II	GTE SPACENET (USA)
1985 Feb. 8	V12	AR3	MARECS B2	INMARSAT
			ARABSAT 1A	ARABSAT
			BRASILSAT S1	EMBRATEL (Brazil)
1985 May 8	V13	AR3	GSTAR I	GTE SPACENET (USA)
			TELECOM 1B	CNES for France TELECOM (Fr.)
July 2	V14	AR1	GIOTTO	ESA
Sept. 12	V15*	AR3	SPACENET III	GTE SPACENET (USA)
1986 Feb. 22	V16	AR1	ECS3	EUTELSAT
			SPOT 1	CNES (France)
			VIKING	CNES for SSC (Swedish Space Corp.-Sweden)
Mar. 28	V17	AR3	GSTAR II	GTE SPACENET (USA)
			BRASILSAT S2	EMBRATEL (Brazil)
May 31	V18*	AR2	INTELSAT V (F14)	INTELSAT

*Launch failure

DATE		FLIGHT	LAUNCH VEHICLE	SATELLITES	CUSTOMER
1987 Sept.	16	V19	AR3	AUSSAT K3	AUSSAT Pty Ltd. (Australia)
				ECS4	ESA for EUTELSAT
1988 Nov.	21	V20	AR2	TV-SAT1	DEUTSCHE BUNDESPOST (FRG)
Mar.	11	V21	AR3	SPACENET IIIR	GTE SPACENET (USA)
				TELECOM 1C	FRANCE TELECOM (France)
May	17	V23	AR2	INTELSAT V (F13)	INTELSAT
June	15	V22/401	AR44LP	METEOSAT P2	ESA
				AMSAT	AMSAT (FRG)
				PAS 1	ALPHA LYRACOM (USA)
July	21	V24	AR3	INSAT 1C	ISRO (India)
				ECS 5	EUTELSAT
Sept.	8	V25	AR3	GSTAR III	GE ASTRO SPACE DIVISION (USA)
				SBS 5	HUGHES AIRCRAFT CORP. (USA)
Oct.	27	V26	AR2	TDF1	CNES for TDF (France)
Dec.	11	V27	AR44LP	SKYNET 4B	Ministry of Defence (Great Britain)
				ASTRA 1A	S.E.S. (Luxemburg)
1989 Jan.	26	V28	AR2	INTELSAT V (F15)	INTELSAT
Mar.	6	V29	AR44LP	JCSAT 1	HUGHES COMM. USA for JC-SAT Japan
				MOP 1	ESA for EUMETSAT
Apr.	1	V30	AR2	TELE-X	SSC (Sweden)
May	26	V31	AR44L	SUPERBIRD A	SPACE COMM. Corp. (Japan)
				DFS KOPERNIKUS 1	DEUTSCHE BUNDESPOST (FRG)
July	11	V32	AR3	OLYMPUS 1	ESA
Aug.	8	V33	AR44LP	TV-SAT 2	DEUTSCHE BUNDESPOST (FRG)
				HIPPARCOS	ESA
Oct.	27	V34	AR44L	INTELSAT VI-(F2)	INTELSAT

APPENDIX 6 *Ariane launch history: the first ten years, 1979–89*

Profile of expenditure for all programmes 1988–2000 (MAU in 1987 e.c.)

Programmes	1988	1989	1990	1991	1992	1993	1994	1995	1996	1997	1998	1999	2000	Total
General budget	124.0	130.2	142.3	152.3	165.4	171.4	174.5	174.5	174.5	174.5	174.5	174.5	174.5	2107.1
Associated to G.B.	57.2	54.8	60.2	65.6	70.0	70.1	71.1	71.5	72.9	74.0	75.1	76.3	77.5	896.3
Science	184.1	193.8	206.0	216.3	226.7	237.9	241.4	240.7	240.6	240.4	240.6	240.6	240.6	2949.7
Other technology progs	4.4	13.0	21.3	24.2	27.6	30.6	32.8	32.8	32.8	32.8	32.8	32.8	32.8	350.7
Earth observation	203.9	234.4	229.9	241.7	247.2	244.6	244.4	237.9	259.9	263.6	259.1	269.0	275.5	3211.1
Microgravity	49.5	82.2	115.6	115.6	115.5	115.5	126.0	136.5	149.1	149.1	149.1	149.1	149.1	1601.9
Telecommunications	251.3	219.1	195.7	232.8	249.0	240.9	273.9	299.9	326.3	346.7	281.4	227.8	253.6	3398.4
Space station and platforms	236.7	325.1	378.9	489.6	499.4	536.2	548.7	580.0	627.1	708.5	711.7	930.1	916.5	7488.5
Space transportation systems	534.1	694.3	864.6	900.0	1017.3	1115.0	1049.1	989.1	842.3	735.3	802.8	622.1	596.9	10762.9
Grand total	1645.2	1946.9	2214.5	2438.1	2618.1	2762.2	2761.9	2762.9	2725.5	2724.9	2727.1	2722.3	2717.0	32766.6

Cumulative expenditure All Programmes

Programmes	1988	1989	1990	1991	1992	1993	1994	1995	1996	1997	1998	1999	2000
General budget	124.0	130.2	142.3	152.3	165.4	171.4	174.5	174.5	174.5	174.5	174.5	174.5	174.5
Associated to G.B.	181.2	185.0	202.5	217.9	235.4	241.5	245.6	246.0	247.4	248.5	249.6	250.8	252.0
Science	365.3	378.8	408.5	434.2	462.1	479.4	487.0	486.7	488.0	488.9	490.2	491.4	492.6
Other technology progs	369.7	391.8	429.8	458.4	489.7	510.0	519.8	519.5	520.8	521.7	523.0	524.2	525.4
Earth observation	573.6	626.2	659.7	700.1	736.9	754.6	764.2	757.4	780.7	785.3	782.1	793.2	800.9
Microgravity	623.1	708.4	775.3	815.7	852.4	870.1	890.2	893.9	929.8	934.4	931.2	942.3	950.0
Telecommunications	874.4	927.5	971.0	1048.5	1101.4	1111.0	1164.1	1193.8	1256.1	1281.1	1212.6	1170.1	1203.6
Space station and platforms	1111.1	1252.6	1349.9	1538.1	1600.8	1647.2	1712.8	1773.8	1883.2	1989.6	1924.3	2100.2	2120.1
Space transportation systems	1645.2	1946.9	2214.5	2438.1	2618.1	2762.2	2761.9	2762.9	2725.5	2724.9	2727.1	2722.3	2717.0

APPENDIX 7   *European Space Agency Budget, 1988–2000*

Appendix 8:

Two excerpts from declarations on European space policy following meeting of European Ministers at The Hague, November 1987

PARTIAL TEXT OF THE RESOLUTION ON THE EUROPEAN LONG-TERM SPACE PLAN AND PROGRAMMES ADOPTED BY THE EUROPEAN SPACE AGENCY COUNCIL MEETING AT MINISTERIAL LEVEL, THE HAGUE, 10 NOVEMBER 1987

The Council, meeting at Ministerial Level...

APPROVES the execution within the Agency of the Ariane 5 launcher development programme starting on 1 January 1988 ... NOTES that the cost of this programme is estimated at 3496 MAU (million accounting units), including three qualification flights;

APPROVES the execution within the Agency of the Columbus development programme according to a phased approach, starting on 1 January 1988 ... under the following conditions:

– the overall cost of this programme as proposed is estimated at 3713 MAU, including the cost of launching the Columbus elements and of the Columbus utilisation preparation:

– Phase 1 of this programme shall consist of a three-year period of initial development within a financial sub-envelope of 669 MAU at the end of which the results of Phase 1 will be reviewed to determine whether the programme objectives can be attained within the overall financial envelope mentioned above ..., taking into account the results of the negotiations with the United States; the decision to embark on Phase 2 related to full development work shall be taken before the end of this three-year period by the Participating States...

– while stressing the desirability in principle of the Polar Mission, studies will be undertaken at the beginning of Phase 1 with the aim of securing user or private sector co-funding for the flight hardware and its operations and to evaluate modified design approaches for ensuring the mission objectives in a cost-effective manner;

APPROVES the execution within the Agency of the Hermes development programme according to a phased approach, starting on 1 January 1988 ... under the following conditions:

– the overall cost of this programme is estimated at 4429.4 MAU, including two qualification flights;

– Phase 1 of this programme shall consist of a three-year period of initial development within a financial sub-envelope of 530 MAU, at the end of which the results of Phase 1 will be reviewed to determine whether the programme objectives can be attained within the overall financial envelope mentioned above; the decision to embark on Phase 2 related to full development work shall be taken before the end of this three-year period by the Participating States...

WELCOMES and ENDORSES the pursuance of the Agency's activities and programmes in the fields of:

– Earth observation, centred around the ERS-1 project already agreed and the proposal for an ERS-2 programme, the continuation of research and development work on a new generation of meteorological satellites, and the preparation for the development of a solid Earth mission and of the Earth observation payloads for the Polar Platform;

– microgravity, with the expansion of microgravity activities in space, centred around an enhanced programme of utilisation of Spacelab, Eureca and new elements of the space transportation and In-Orbit Infrastructure;

– space telecommunications, with the undertaking of the payload and spacecraft development and experimentation programme (PSDE) and the continuation of an advanced systems and technology programme;

ENDORSES the decision to carry out the long-term scientific programme on the basis of the 'Horizon 2000' programme, and as a first step towards undertaking the Solar Terrestrial Science Programme.

RECALLING that it agreed, in order to re-inforce space science activities in Europe over the next decade, progressively to increase the level of funding of the mandatory science programme to reach 177 MAU by 1989, AGREES progressively to increase the science programme budget to reach a level of 216.7 MAU at the end of the next period of the level of resources (1992).

PARTIAL TEXT OF THE RESOLUTION ON PARTICIPATION IN THE SPACE STATION PROGRAMME ADOPTED BY THE EUROPEAN SPACE AGENCY COUNCIL MEETING AT MINISTERIAL LEVEL, THE HAGUE, 10 NOVEMBER 1987

The Council, meeting at Ministerial Level...

CONFIRMS its interest in participating with the United States in the international Space Station, provided that ... as a minimum the following essentials are met:

(1) that the responsibility of ESA in the design, development, operation and utilisation of the elements Europe will provide be recognised in the Intergovernmental Agreement and the Memorandum of Understanding, while accepting NASA's overall programme co-ordination and direction regarding the manned base. ESA will remain in full control of the design, development, operation and utilisation of the Man-Tended Free Flyer and

of the European Polar Platform while respecting jointly agreed NASA/ ESA technical interface documentation and overall Space Station safety requirements;

(2) that, whereas the present draft Intergovernmental Agreement's provisions concerning the civil nature and the peaceful purposes of the Space Station and each Partner's right to determine the application of these provisions to its elements are considered adequate and their substance must not be compromised, a satisfactory solution in line with the ESA Convention be found to the problem raised by the United States' requirement for recognition by the other Partners of the United States' right to interpret the notion of peaceful purposes in relation to the US utilisation of the elements they provide;

(3) that adequate provisions be made for settlement of disputes concerning, as the case may be, the interpretation or implementation of the Intergovernmental Agreement or the Memorandum of Understanding;

(4) that a legal regime safeguarding the interests of the Participating States and of European users be achieved;

AGREES, in the event that the negotiators do not succeed in fulfilling the above-mentioned conditions within a time frame compatible with the Columbus Development Programme,

– to adapt the content of the Columbus Development Programme ... to the new situation, as quickly as possible and in line with the objectives at the Rome Ministerial Council meeting ... of 31 January 1985, while seeking other forms of co-operation with the United States regarding the respective manned spaceflight programmes,

– to ensure also in this event coherence between the elements of the European In-Orbit Infrastructure and to strive to maintain, in line with the financial commitment of Member States concerned, the present level of their responsibilities and industrial involvement in the Development Programme.

Appendix 9

LANDMARKS IN SPACE HISTORY

1957 Oct.   4– Soviet Union launches Sputnik 1, world's first artificial satellite
1961 Apr. 12– Soviet cosmonaut Yuri Gagarin makes first manned spaceflight aboard Vostok 1
1962 Feb. 20– Astronaut John Glenn becomes first American in space aboard Friendship 7
1962 Mar. 29– Signature of ELDO convention, London
1964 Feb. 29– ELDO convention comes into force
1964 Mar. 20– ESRO convention comes into force
1965 Nov. 26– France joins élite club of space nations by launching satellite on Diamant rocket
1966 Dec. 13– European Space Conference holds first meeting, Paris
1967 Apr.   – Publication of Bannier report recommending re-organisation of ESRO
1967 May 29– Launch failure of ESRO's first satellite, ESRO-2A
1968 May 17– Successful launch of ESRO satellite, ESRO-2B
1969 Jul. 20– Apollo 11 lands on moon
1971 Nov.   5– Europa II launch failure
1973 July   – European Space Conference gives go-ahead for Spacelab and Ariane programmes and foundation of European Space Agency
1975 May 31– ESA starts operations
1977 Sept. 13– Launch failure of OTS test satellite
1978 May 11– Successful launch of replacement OTS-2
1979 Dec. 24– Successful first flight of Ariane-1 rocket
1980 May 23– Failure of second Ariane-1 flight
1981 Apr. 12– US shuttle makes first flight
1982 Jun. 24– French astronaut Jean-Loup Chrétien becomes first West European to fly in space, aboard Soviet Soyuz T-6 mission to Salyut space station
1982 Sept. 10– Failure of fifth Ariane-1 flight
1983 Jun. 16– ECS-1 (first fully-operational European telecommunications satellite) launched
1983 Nov. 28– Launch of first Spacelab flight on shuttle Columbia with West German ESA astronaut Ulf Merbold aboard
1984 Aug.   4– First flight of upgraded Ariane-3 rocket

214

1985 Jan. 30– Two-day ESA ministerial council meeting starts in Rome, laying groundwork for future Columbus, Hermes and Ariane-5 manned space programme

1985 Jun. 17– French astronaut Patrick Baudry launched on mission aboard US shuttle Discovery

1985 Sept. 12– Failure of Ariane-3 flight 15

1985 Oct. 30– Dutch ESA astronaut Wubbo Ockels and West German astronauts Reinhard Furrer and Ernst Messerschmid launched on Spacelab mission aboard Challenger

1986 Jan. 28– Space shuttle Challenger explodes during launch, killing all seven crew and grounding shuttle programme

1986 Mar. 14– European Giotto probe flies within 600 km of Halley's Comet nucleus

1986 Mar. 28– First Ariane launch from new ELA-2 launchpad, Kourou

1986 May 31– Failure of Ariane-2 flight 18, leading to 16-month launch interruption

1987 Jul. 23– British Prime Minister Margaret Thatcher rules out increasing Britain's space budget

1987 Aug. 4– Roy Gibson, head of British National Space Centre, resigns

1987 Sept. 16– Ariane launches successfully resume

1987 Nov. 9– Two-day meeting starts of European ministers which gives go-ahead for Columbus, Hermes and Ariane-5 manned space programme

1988 Jun. 15– First flight of upgraded Ariane-4 rocket

Notes

Chapter 2 The Pioneering Years

1.   *Le Grand Atlas de l'Espace* (Paris: Universalis, 1987) p. 28.
2.   *Le Grand Atlas de l'Espace* (Paris: Universalis, 1987) p. 36.
3.   T. Bower, *The Paperclip Conspiracy* (London: Michael Joseph, 1987) pp. 58–61.
4.   *Le Grand Atlas de l'Espace* (Paris: Universalis, 1987) p. 30.
5.   T. Bower, *The Paperclip Conspiracy* (London: Michael Joseph, 1987) p. 145.
6.   '30 years of space exploration', Spaceflight, Vol. 29 :October 1987).
7.   P. Wright, *Spycatcher* (New York: Viking Penguin, 1987) p. 26.
8.   Lord Rippon of Hexham, speaking in House of Lords debate on space, 30 March 1988, *Hansard*, Vol. 495, No. 106, p. 799.
9.   E. Amaldi, 'Why we need a European Organisation for space research' in *Europe – Two decades in Space* (Noordwijk, the Netherlands: ESA, 1984) p. 10.
10. A. Dupas, 'The history of space activities in Europe' in *1986 European Space Directory* (Paris: Sevig, 1986) p. 30.
11. ELDO 1960–65 report to Council of Europe, p. 9.
12. ELDO 1960–65 report to Council of Europe, p. 5.
13. Report by Société pour l'Etude et la Réalisation d'Engins Balistiques (SEREB) to French Defence Ministry, 23 December 1960. Quoted by French space agency Centre National d'Etudes Spatiales (CNES) in 'Les Lanceurs Diamant', *25 ans d'Espace en France* (Paris: CNES, 1987).

Chapter 3 The Europa Débâcle

1.   R. Aubinière, 'Tenth anniversary of the Establishment of ELDO', ESRO/ELDO Bulletin No 24 (March 1974) p. 12.
2.   Lord Rippon of Hexham, House of Lords debate on space policy, *Hansard*, vol. 495, no. 106 (30 March 1988) p. 800.
3.   ESRO General Report 1969, 'Towards a European Space Policy' (1969) p. 10.
4.   Twenty years of European co-operation in space 1964/84 (Noordwijk, the Netherlands: ESA, 1984) p. 119.
5.   P. Wright, *The Times* (6 November 1971).
6.   R. Aubinière, 'Tenth anniversary of the Establishment of ELDO', ESRO/ELDO Bulletin No. 24 (March 1974) p. 13.
7.   R. Aubinière, 'Tenth anniversary of the Establishment of ELDO', ESRO/ELDO Bulletin No. 24 (March 1974) p. 14.

Chapter 4 Ariane – From Drawing-Board to Launchpad

1.   R. Vignelles and P. Rasse, 'The Ariane launcher and its progress', ESA Bulletin No. 15 (August 1978) p. 10.
2.. 'Projects under development', ESA Bulletin No. 11 (December 1977) p. 47.
3.   R. Orye, 'Introduction', ESA Bulletin No. 15 (August 1978) p. 8.
4.   P. Creola, 'Ariane – the road to independence', in *Europe – Two decades in space 1964/84* (Noordwijk, the Netherlands: ESA, 1984) p. 33.

5.   P. Creola, 'Ariane – the road to independence', ibid., p. 34.
6.   M.-A. Hauzeur, 'Operational qualification programme of Guiana space centre' in ESA Bulletin No. 20 (November 1979) p. 43.
7.   P. Creola, 'Ariane – the road to independence', ibid., p. 36.

Chapter 5   Early Scientific Satellites

1.   ESRO General Report 1966 (Paris: ESRO, 1966) p. 19.
2.   H. Bondi, 'ESRO in 1967–68' in *Europe – Two decades in space 1964/84* (Noordwijk, the Netherlands: ESA, 1984) p. 21.
3.   ESRO General Report 1967 (Paris: ESRO, 1967) p. 106.
4.   Bannier report (1967) cited by H. Bondi, 'ESRO in 1967–68', op. cit., p. 22.
5.   D. E. Page, 'The success of ESA's eight scientific satellites', ESA Bulletin No. 3 (October 1975) p. 13.
6.   Ibid., p. 14.
7.   *Twenty years of European space cooperation 1964/84* (Noordwijk, the Netherlands: ESA, 1984) p. 34.
8.   K.-P. Wenzel, 'Last orbit for HEOS-1, ESA's longest-living satellite', ESA Bulletin No. 4 (February 1976) p. 14.
9.   ESRO 1968 annual report (Paris: ESRO, 1968) p. 7.
10.   A. Hocker, 'The rescue', in *Europe – Two decades in space 1964/84* (Noordwijk, the Netherlands: ESA, 1984) p. 64.
11.   ESRO 1972 annual report (Paris: ESRO, 1972) p. 7.
12.   R. Gibson, 'Introduction', ESA Bulletin No. 2 (August 1975) p. 4.
13.   'Geos launch failure and rescue operation', ESA Bulletin No. 9 (May 1977) p. 2.

Chapter 6   Early Communications and Weather Satellites

1.   E. S. Mallett, 'Why Europe must invest in communications satellite systems', ESA Bulletin No. 14 (May 1978) p. 5.
2.   Interview with Andrea Caruso, Paris, 9 December 1988.
3.   Idem.
4.   A. Lebeau, 'The changing role of Europe in space', ESA Bulletin No. 6 (August 1976) p. 4.
5.   'Ten years of Meteosat pictures', speech by P. Goldsmith, Darmstadt, West Germany, 3 December 1987.
6.   K. H. Stewart, 'The significance of Meteosat for meteorology', ESA Bulletin No. 11 (December 1977) p. 9.

Chapter 7   Ariane – The Commercial Challenge

1.   P. Creola, 'Ariane – the road to independence', in *Europe – Two Decades in Space 1964–84* (Noordwijk, the Netherlands: ESA, 1984) p. 36.
2.   Programmes under development and operations, ESA bulletin No. 23 (August 1980) p. 47.
3.   I. Gravière, 'Ariane sur le fil', *Le Nouveau Journal* (17 June 1983) p. 25.
4.   Frédéric D'Allest speaking at Kourou launch site, 30 May 1986.
5.   D. E. Fink, 'NASA after Challenger', *Aviation Week and Space Technology*, Vol. 124, No. 23 (9 June 1986) p. 11.
6.   Michel Lavany, Arianespace press briefing, Paris, 9 January 1987.

7. C.-M. Vadrot, 'Chirac innocente les espions d'Ariane', *Le Matin*, Paris (14 May 1987).
8. G. Marion, 'L'affaire du réseau Ariane – L'épouse sovietique de M. Pierre Verdier pourrait beneficier rapidement d'un non-lieu', *Le Monde*, Paris (15 May 1987).
9. G. Marion, 'L'affaire du réseau Ariane – L'épouse soviétique de M. Pierre Verdier pourrait bénéficier rapidement d'un non-lieu', *Le Monde*, Paris (15 May 1987).
10. Frédéric d'Allest, press conference at Bordeaux Technospace exhibition, 2 December 1986.
11. J. N. Wilford, 'Ariane delay expected to strain Western rocketry', *New York Times* Service, reprinted in *International Herald Tribune*, Paris (4 December 1986).
12. G. Dupuy, 'Un challenge à trois étages', *Libération*, Paris (15 September 1987).
13. J.-P. Croize, 'Ariane – un lancement crucial pour l'Europe spatiale', *Le Figaro*, Paris (15 September 1987).
14. Frédéric D'Allest briefing journalists, Kourou, 16 September 1987. Quoted in Associated Press report, *International Herald Tribune*, Paris (17 September 1987).
15. Frédéric D'Allest, 'Ariane back to service', Arianespace newsletter (October 1987).
16. J.-P., Croize, 'Un prodigieux puzzle', *Le Figaro*, Paris (15 June 1988).
17. Press conference, ESA headquarters, Paris, 15 June 1988.
18. Michel Lavany, Arianespace sales manager Latin America, Asia and Italy, press briefing in Paris, 9 January 1987.
19. J. M. Lenorowitz, 'Europe presses US to agree on launch competition rules', *Aviation Week and Space Technology*, Vol. 128 No. 26 (27 June 1988).

Chapter 8   Spacelab

1. Jean-Jacques Dordain, ESA, speaking at Technospace symposium, Bordeaux, France, 8 December 1988.
2. A. Thirkettle, F. Di Mauro and R. Stephens, 'Spacelab – from early integration to first flight: part two', ESA bulletin No. 39 (August 1984) p. 73.
3. Ibid., p. 74.
4. E. Quistgaard, 'Spacelab – Triumphant conclusion to a decade of cooperation', ESA Bulletin No. 37 (February 1984) p. 14.
5. Jean-Jacques Dordain, ESA, speaking at Technospace symposium, Bordeaux, France, 8 December 1988.
6. E. Quistgaard, 'Spacelab – Triumphant conclusion to a decade of cooperation', ESA Bulletin No. 37 (February 1984) p. 15.
7. Wubbo Ockels, Spacelab D-1 payload specialist, speaking at Technospace symposium, Bordeaux, France, 8 December 1988.
8. E. Quistgaard, 'Spacelab – Triumphant conclusion to a decade of cooperation', ESA Bulletin No. 37 (February 1984) p. 15.
9. C. Covault, 'Spacelab 3 mission demonstrates crystal growth, animal care challenges', *Aviation Week and Space Technology*, Vol. 122 (6 May 1985) p. 20.
10. Ibid., p. 18.
11. Wubbo Ockels, Spacelab D-1 payload specialist, speaking at Technospace symposium, Bordeaux, France, 8 December 1988.

12. Prof. Heinz Wolff, chairman of ESA Microgravity Advisory Committee, giving evidence to House of Lords Select Committee on Science and Technology, Subcommittee 1, hearings on UK space policy, 4 March 1987 (Session 1987–88, 2nd report, p. 28).
13. Wubbo Okels, Spacelab D-1 payload specialist, speaking at Technospace symposium, Bordeaux, France, 8 December 1988.
14. T. M. Foley (quoting Arthur Davidsen), 'NASA will cancel many spacelab missions', *Aviation Week and Space Technology*, Vol. 125 (1 September 1986) p. 40.

Chapter 9   Remote Sensing, Comets and the Lure of Mars

1. Claude Martinand, director-general of France's Institut Géographique National (IGN), speaking at a Paris press conference, 27 November 1987.
2. Jean-Pierre Contzen, director-general, European Commission Joint Research Centre, speaking at Space Commerce '88 symposium, Montreux, Switzerland, 23 February 1988.
3. Jean-Pierre Contzen speaking at Space Commerce '88 symposium, Montreux, Switzerland, 23 February 1988.
4. Michel Courtois, CNES official, speaking at Paris press conference, 27 November 1987.
5. Thomas Pyke, NOAA assistant administrator for Satellite and Information Services, speaking at Space Commerce '88 symposium, Montreux, Switzerland, 23 February 1988.
6. Gérard Brachet, Spot Image President, speaking at Paris press conference, 27 November 1987.
7. Anatoliy Nicolaev, Director Soyuzkarta, speaking at Space Commerce '88 symposium, Montreux, Switzerland, 23 February 1988.
8. Yulian Novikov, Chief of Department, USSR Ecological Committee, speaking at Space Commerce '88 symposium, Montreux, Switzerland, 23 February 1988.
9. Charles Williams, Eosat president, speaking at Space Commerce '88 symposium, Montreux, Switzerland, 23 February 1988.
10. Don Walklet, president, Terra-Mar Resource Information Services, speaking at Space Commerce '88 symposium, Montreux, Switzerland, 23 February 1988.
11. R. Bonnet quoted in J. M. Lenorovitz, 'Giotto redirected to fly past Earth after returning data on Halley's', *Aviation Week and Space Technology*, Vol. 124 (24 March 1986) p. 22.
12. Objectif Phobos, CNES document, June 1988.
13. 'L'invasione di Marte comincia da Phobos', *La Repubblica*, Rome (28 March 1989) p. 10.
14. Memorandum by the British Meteorological Office, written evidence to the House of Lords Select Committee on Science and Technology, Sub-committee 1, on UK space policy (Session 1987–88 2nd report) pp. 309/10.
15. P. Goldsmith, speech entitled 'Ten years of Meteosat pictures', Darmstadt, West Germany, 3 December 1987.
16. B. Mason quoted in M. Degli Esposti, 'Al via super-Meteosat, sentinella dell'ambiente', *Il Sole 24 Ore*, Milan (21 March 1989).

Chapter 10   The Telecommunication Satellite Revolution

1. Interview with Andrea Caruso, Eutelsat director-general, Paris, 9 December 1988.

2. Interview with Claude Goumy, director-general, Matra Defence and Space Division, Paris, 8 March 1988.
3. Ibid.
4. Marcel Roulet, director-general France Télécom, speaking after Kourou launch of Telecom IC satellite, 11 March 1988.
5. Gerry Waylan, President of GTE Spacenet, speaking after Kourou launch of Spacenet IIIR satellite, 11 March 1988.
6. Air Vice-Marshal E. H. Macey, Assistant Chief of British Defence Staff (Policy and Nuclear), testifying to House of Lords Select Committee on Science and Technology, Subcommittee 1, on UK space policy, 28 October 1987 (Session 1987–88, 2nd report, p. 157).
7. British Ministry of Defence memorandum, written evidence to House of Lords Select Committee on Science and Technology, Subcommittee 1 on UK space policy (Session 1987–88, 2nd report, p. 147).

Chapter 11   Satellite Television

1. R. Snoddy, 'Programme of expansion', *Financial Times* 'Survey on Satellite Broadcasting' (29 July 1988), p. I.
2. Ibid.
3. Herman Strub, head of space division, West German Ministry of Research and Technology (BMFT), speaking at Munich press conference, 9 July 1986.
4. R. Mihail, 'Satellites de télévision: la panne', *Le Point*, Paris, No. 722 (21 July 1986) pp. 48–9.
5. French government spokesman Denis Baudouin, speaking at press briefing, Paris, 29 July 1986.
6. Alain Juppé letter to Jacques Chirac, quoted in *Le Figaro*, Paris (3 February 1987).
7. A. Ducrocq, 'Feu TDF-1', *Air & Cosmos,* Paris No. 1130, (14 February 1987).
8. Interview with Andrea Caruso, Eutelsat director-general, Paris, 9 December 1988.
9. Ibid.
10. Interview with Steve Maine, Head of Broadcast and Visual Services, British Telecom International, London, 10 February 1989.
11. Sky television advertisement in *Sunday Times*, London, 15 January 1989.
12. M. Lawson, 'The moronic inferno', *Independent*, London (7 February 1989).
13. Anthony Simonds-Gooding, BSB chief executive, quoted in R. Evans, 'TV aerial to woo satellite viewers', *The Times*, London (3 August 1988).
14. Georg-Michael Luyken, deputy director of The European Institute for the Media, Manchester University and executive secretary of European television committee, quoted in R. Snoddy, 'Satellite public service TV channel urged for Europe', *Financial Times*, London (26 May 1988).

Chapter 12   The French Manned Space Programme

1. Report of the French Académie des Sciences, 'La Recherche et la Politique Spatiale dans les prochaines décennies', 21 March 1988.
2. Jacques Chirac, speech to opening ceremony of Le Bourget air show, Paris, 10 June 1987.
3. Jean-Loup Chrétien, interviewed from Baikonour on French Antenne 2 television news, 26 November 1988.

4. Philippe Couillard, director of manned space flight, CNES, speaking from Baikonour on French Antenne 2 television, 26 November 1988.
5. Jean-Loup Chrétien, interviewed from Mir on French Antenne 2 television news, 4 December 1988.

Chapter 13 Satellites and Probes in the 1990s

1. S. Ride, Leadership and America's Future in Space (report to NASA, August 1987).
2. Livio Marelli, head of ESA's Earthnet programme office, speaking at briefing at ESRIN, Frascati, Italy, 22 March 1988.
3. Luigi Fusco, Research and Support officer, ESA's Earthnet programme office, speaking at briefing at ESRIN, Frascati, Italy, 22 March 1988.
4. Roberto Somma, Selenia Spazio deputy director marketing with responsibility for Meteosat project, quoted in M. Degli Esposti, 'Al via super-Meteosat sentinella dell'ambiente', *Il Sole 24 Ore* (21 March 1989).
5. NASA Topex-Poseidon programme official Bill Townsend speaking at press conference, Paris, 6 May 1988.
6. NASA Topex-Poseidon programme official Stanley Wilson speaking at press conference, Paris, 6 May 1988.
7. 'Allied participation in SDI research', Report to the Congress on the Strategic Defense Initiative (Washington DC: April 1987).
8. J. Isnard, 'Le satellite militaire français Helios pourra espionner des radars adverses', *Le Monde*, Paris (25 March 1987).

Chapter 14 Shuttles, Space Stations and Politics

1. Resolution adopted by ESA Council meeting at ministerial level, Rome, 31 January 1985.
2. 'Forward to the future', European Space Agency document (Paris: 1985).
3. Ernst Messerschmid speaking to 'Europe in Space – the manned space system' symposium, Strasbourg, France, 29 April 1988.
4. J.-E. Feustel-Buechl, director of space transport systems, ESA, speaking at Technospace symposium, Bordeaux, France, 7 December 1988.
5. Resolution adopted by ESA Council meeting at ministerial level, Rome, 31 January 1985.
6. CNES statement, 18 October 1985.
7. P. Langereux, 'Hermes: Aérospatiale et Dassault se partagent la maitrise d'oeuvre', *Air & Cosmos*, Paris, No. 1067 (26 October 1985) p. 41.
8. CNES director-general Frédéric d'Allest, quoted in P. Langereux, 'Hermes: Aérospatiale et Dassault se partagent la maîtrise d'oeuvre', *Air & Cosmos*, Paris, No. 1067 (26 October 1985) p. 44.
9. D. Pavy, 'Paris pense que Bonn ne pourra pas fuir longtemps ses responsabilités', *Les Echos*, Paris (9 January 1986) p. 8.
10. See H. Mark, *The Space Station, a personal journey* (Durham, North Carolina: Duke University Press, 1987).
11. President Reagan, State of the Union address to Congress, 25 January 1984.
12. Joint statement following meeting between US, Japan, Canada and European countries on space station, Washington DC, 12 February 1987.
13. BNSC director-general Roy Gibson, quoted in 'NASA changes tune', *Flight International* (28 February 1987).
14. ESA director-general Reimar Luest speaking at press conference, Paris, 19 May 1987.

15. Richard Halpern, NASA director of space station utilisation, quoted in P.
     Marsh, 'US ready to build space station alone', *Financial Times* (9 October
     1987) p. 5.
16. R. Gibson 'Britain and Space', J. D. Bernal lecture, 1987.
17. British Prime Minister Margaret Thatcher, House of Commons, Oral
     Answers, *Hansard* (23 July 1987) p. 480.
18. Sir Geoffrey Pattie quoted in M. Cross, 'Space chief resigns over lack of UK
     commitment', *Independent*, London, 5 August 1987.
19. Reimar Luest, ESA director-general, telephone interview in London, 5
     August 1987.
20. Kenneth Clarke, British Trade Minister, House of Commons, Oral
     Answers, *Hansard* (28 October 1987) p. 292.
21. Kenneth Clarke, 'Putting space to work', *The Times*, London (4 November
     1987).
22. 'Out of space', *The Times*, London (9 November 1987).
23. 'Europe's aims in space', *Financial Times*, London (9 November 1987).
24. Kenneth Clarke, speech to ESA Council meeting at ministerial level, The
     Hague, 9 November 1987.
25. Reimar Luest, interview following ESA ministerial meeting, The Hague, 10
     November 1987.
26. Heinz Riesenhuber, West German Research and Technology Minister,
     speaking at press conference after ESA ministerial meeting, The Hague, 10
     November 1987.
27. French Industry Minister Alain Madelin, speaking at press briefing after
     ESA
     ministerial meeting, The Hague, 10 November 1987.
28. Kenneth Clarke, speaking at press conference after ESA ministerial meet-
     ing, The Hague, 10 November 1987.
29. Jacques-Louis Lyons, president of CNES, briefing journalists after ESA
     ministerial meeting, The Hague, 10 November 1987.
30. Joseph Allen, executive vice-president Space Industries Inc, speaking at
     Space Commerce '88 symposium, Montreux, Switzerland, 24 February 1988.
31. Fredrik Engstrom, head of ESA's space station directorate, speaking at
     Space Commerce '88 symposium, Montreux, Switzerland, 24 February 1988.
32. Jean-Jacques Dordain, head of ESA's space station and platforms promo-
     tion and utilisation department, speaking at Space Commerce '88 sympo-
     sium, Montreux, Switzerland, 24 February 1988.
33. Wubbo Ockels, speaking at Technospace symposium, Bordeaux, France, 8
     December 1988.
34. Claude Nicollier, ESA astronaut, speaking at 'Europe in Space – the
     manned space system' symposium, Strasbourg, France, 29 April 1988.
35. Roy Gibson, testifying before House of Lords Select Committee on Science
     and Technology (Sub-committee 1), 28 October 1987 (Session 1987–88, 2nd
     report) Vol. 2, p. 167.
36. Roy Gibson, testifying before House of Lords Select Committee on Science
     and Technology (Sub-committee 1), 8 July 1987 (Session 1987–88, 2nd
     report) Vol. 2, p. 90.
37. S. W. Korthals-Altes, 'Will the Aerospace Plane Work?', *Technology Re-
     view*, edited at the Massachusetts Institute of Technology (2 January 1987).
38. Lou Harrington, McDonnell Douglas Corp senior vice-president, speaking
     at aerospace symposium in Paris on eve of Le Bourget air show, 9 June 1987.
39. Sir Geoffrey Pattie, 'Youth left on the launch pad', *Sunday Times*, London
     (31 July 1988).

40. J. A. Holt, British Aerospace managing director (space and communications division), testifying to House of Lords Select Committee on Science and Technology (Sub-committee 1), 11 November 1987 (Session 1987–88, 2nd report) Vol. 2, p. 183.

41. Sir Geoffrie Pattie, testifying to House of Lords Select Committee on Science and Technology (Sub-committee 1), 25 November 1987 (Session 1987–88, 2nd report) Vol. 2, p. 212.

42. See J. Partiot, 'Space Insurance, a crisis situation?', *Aérospatiale* magazine, Paris, January 1986.

43. Interview with Andrea Caruso, Eutelsat director-general, Paris, 9 December 1988.

44. Andrea Caruso, Eutelsat director-general, speaking at Space Commerce '88 symposium, Montreux, Switzerland, 24 February 1988.

45. Allan McCaskill, manager, Intelsat launch vehicle programmes, speaking at Space Commerce '88 symposium, Montreux, Switzerland, 24 February 1988.

Chapter 15   The International Scene and the Way Forward

1. S. Ride, 'Leadership and America's future in space', a report to the NASA administrator, August 1987.

2. Report of the Presidential Commission on the Space Shuttle Challenger Accident, Washington DC (6 June 1986).

3. 'Liftoff, liftoff', *Newsweek*, New York (10 October 1988) p. 12.

4. J. Kifner, 'Israel is Eighth Nation to Launch a Satellite', *New York Times Service* article printed in *International Herald Tribune*, Paris (20 September 1988).

5. Lord Young of Graffham, House of Lords debate on space policy, *Hansard*, Vol. 495, No. 106 (30 March 1988) pp. 826–9.

6. 'United Kingdom Space Policy', Vol. 1, report of House of Lords Select Committee on Science and Technology, Session 1987–88, 2nd report (17 December 1987) pp. 61–2.

Index

A1-4 rockets, 4–5
Académie des Sciences (French
    Academy of Sciences), 134
Advertising, 136
AEG-Telefunken, 43, 70, 95, 112
Aeritalia, 43, 70, 145, 158, 172
Aérospatiale, 23, 29, 40, 46, 63, 101,
    112, 131, 137, 146, 156–8
AFP (Agence France Presse), 63
Air & Cosmos, 115
Airbus, 157
Alcatel Espace, 103, 107, 112
Aldrin, Buzz, 17
Allen, Joseph, 172
Al-Saud, Sultan Abdul Aziz, 131–2
Amaldi, Edoardo, 8
Amar, Paul, 56
Ames Research Center, 82
AMI (Active Microwave
    Instrument), 142
AMSAT, 26, 54, 66
Amstrad, 118
Andoya, 189
Animals in space, 81– 2
ANT Nachrichtentechnik, 101, 107,
    112, 158
APEX (Ariane Passenger
    Experiments), 25
APM (Attached Pressurised
    Module), 172–3
    see Columbus
Apollo, 1, 17, 19, 70, 124, 127–8
APPLE, 26, 52
Arabsat, 55, 131
Arab Satellite Communications
    Organisation, 131
Archangel, 124
Arets, Jean, 162
Ariane, 22, 23–32, 49–69, 74, 87, 95,
    100, 102, 104–6, 108, 113, 116,
    121, 128, 144, 154–7, 163, 165,
    167–71, 183– 4, 186–8, 194–6
Arianespace, 54, 65, 67–9, 154, 186,
    189– 90, 199
Armstrong, Neil, 17
Ascension Island tracking station,
    31–2, 53
As de Coeur, 138

Astérix, 14
Astra, 67, 113, 117–22
Astris, 16
Astronauts, European, 77, 83–5,
    123, 125–39, 155
    see Baudry
        Chrétien
        Furrer
        Messerschmid
        Ockels
        Tognini
Astronomy, 34, 37–41, 78, 86, 133,
    145–6
    see COS- B
        Cosmic radiation
        Gamma-ray astronomy
        Hipparcos
        ISA
        LAS
        Solar observation
        Solar radiation
        Spartan
        TD-1/2
        Ultra-violet astronomy
        X-ray astronomy
Astrophysics, 86, 129–30, 135
Atlantis, 85
Atlas Centaur, 26, 68, 188, 196
Atmospheric research, 6–7, 33– 41,
    45–8, 100–1, 123–4
    see International Geophysical
        Year
        Meteorological satellites
AT&T (American Telephone and
    Telegraph), 131
Aubinière, General R., 15, 20, 21
Auger, Pierre, 8, 34, 35
Auréol, 124
Aurora Borealis, 33, 36, 38
Aussat, 65
Australia, 9, 65, 196
    see Aussat
        Cape York
        Woomera
Austria, 71, 98, 140, 170
*Aviation Week and Space
    Technology*, 60
Avions Marcel Dassault, 156–8

B-52, 178
Baikonour, 123, 125, 128, 131, 136
Bannier, J. H.,
  see Bannier report
Banier report, 34–5
Baudouin, Denis, 114
Baudry, Patrick, 126–33, 135, 138,
  174, 176
Bayh amendment, 150
BBC (British Broadcasting
  Corporation), 151
Belgium, 9, 16–17, 24, 40, 46, 71, 87,
  111, 158, 170
Berezovoi, Anatoli, 128
Berry, William, 82
Big Bird, 60
Blue Streak, 7, 8, 9, 11, 15, 16, 18,
  183–4
Bluford, Guion, 83–4
BMFT (West German Research and
  Technology Ministry), 112, 153,
  159, 179
BNSC (British National Space
  Centre), 161, 163–5, 177, 183
Bock, Professor G., 11
Bond, Alan, 119, 181
  see Bond Corporation
Bond Corporation, 119
Bondi, Sir Hermann, 18, 34, 35
Bonnet, Roger, 96
Bourdiol, Pierre-Antoine, 63
Brachet, Gerard, 92
Brandenstein, Charles, 133
Brandt, Willy, 20
Brasilsat, 55, 58
Brazil, 55, 58
  see Brasilsat
Brezhnev, Leonid, 129
Britain, 2, 7–9,11–12, 15–16, 20, 40,
  43, 45, 46, 71, 82, 95, 100–1,
  105–7, 111, 118–20, 122, 140,
  146, 149–52, 163–71, 176,
  179–84, 191, 197–9
  see Blue Streak
    Skylark
    Skynet
British Aerospace, 40, 71, 95, 106,
  109, 146, 158, 168, 179, 182–4
British Telecom, 118
BSB (British Satellite Broadcasting),
  119–20
Buchli, James, 83–4
Bulgaria, 98, 127

Buran, 136, 175, 193
Bush, George, 150, 192

Campbell, Duncan, 151
Canada, 154, 161
Candide, M. S., 39
Cape York, 196
Cape Canaveral, 1, 41, 74–5, 81, 131
Cargus, 179
  see Saenger
Carimare, 29
Carroll Group, 184
Caruso, Andrea, 45, 102, 117–18,
  187
CASA (Construcciones
  Aeronauticas SA), 40, 158
CASDN (Comité d'Action
  Scientifique de la Défense
  Nationale), 6
Cassini, 147
CAT satellite, 32
CEPT (European Conference of
  Postal and Telecommunications
  Administrations), 42, 45
CERN (European Centre for
  Nuclear Research), 8, 43
CETS (European Conference on
  Telecommunications by
  Satellite), 42
Challenger, 57, 60, 61, 67, 81–3,
  85–6, 105–6, 134, 156, 173,
  175–6, 187–8, 192–3, 195
Chernobyl, 87–8, 141
China, 68–9, 188, 193, 195–7
  see Long March
Chirac, Jacques, 62, 64, 114–16,
  134
Chrétien, Jean- Loup, 77, 123,
  125–39, 174
CIT Research, 112
Clarke, Kenneth, 165–6, 168,
  170–1
CLUSTER, 148
  see SOHO
CNN (Cable News Network), 111
CNES (Centre National d'Etudes
  Spatiales), 13–14, 20–1, 22, 23,
  27, 38, 46, 54, 87, 91–2, 98, 112,
  126, 134, 137, 144, 157–8
CNS (Italian national research
  institute), 48
Collins, Michael, 17
Columbia, 75–7, 79–81, 85

Columbus, 147, 153–4, 156, 159, 163–75, 185
  see International Space Station
Comsat General, 108
Comtesse du Barry, 139
Concorde, 56, 136, 179–80
Congress, US, 151, 159, 162
Conrad, Charles 'Pete', 73
Contamine, Claude, 115–16
Contzen, Jean-Pierre, 90
COPERS (European Preparatory Conference for Space Research), 8–9
Coralie, 9, 16
COSARI, 26
COS-B, 26, 39–41
Cosmic radiation, 36, 37, 39–41, 125, 138
Cosmos, 92, 125
COSPAR (International Council of Scientific Unions' Astrophysics Committee), 8
Couillard, Philippe, 137
Courtois, Michel, 91
Creola, Peter, 28, 32, 51
Cryorocket, 18
CTS (Canadian Telecommunications Satellite), 26
Cuba, 127
Curien, Hubert, 67
Cytos, 125
Czechoslovakia, 98, 127

D1 Spacelab mission, 77, 83–5
  see Spacelab
D2-MAC Packet, 117, 119
D'Allest, Frédéric, 54, 56, 59–60, 62, 64–7, 69, 158
Darmstadt
  see ESDAC
     ESOC
DAS, 98
Davidsen, Arthur, 86
Davis, Bob, 199
DBS (Direct Broadcast Satellite), 111–12, 115–16, 121
Defence Ministry, British,
  see Ministry of Defence, British,
Defense Department, US,
  see Pentagon
De Gaulle, Charles, 49, 123
De Havilland, 7, 8
Deimos, 97

Delta, 44, 46, 47, 60, 61, 68, 187–8, 194, 196
Demark, 10, 40, 46, 71, 170
Devil's Island, 49
DFS Kopernikus, 107
DFVLR (Deutsche Forschungs- und Versuchsanstalt fuer Luft- und Raumfahrt), 112
Diamant, 13–14, 54
Discovery, 130–3, 173
Disney, 119
DMA (French arms procurement agency), 13–14
Dordain, Jean-Jacques, 74, 174
Dornberger, General Walter, 5
Dornier, 70, 82, 95, 148, 158
Dragon, 124
DRS (Data Relay Satellite), 146–7
DST (French counter-intelligence service), 62, 63
Ducrocq, Albert, 115
Dunbar, Bonnie, 83–4
Dzhanibekov, Vladimir Alexandrovich, 123, 125, 129–30, 133

Early Bird, 42, 107
Earth observation, 57, 78, 80, 85, 87–94, 101, 135, 141–4, 151, 168, 172, 191, 194
  see Eosat
     ERS
     Landsat
     MOS-1
     Polar platform
     Pollution
     Spacelab
     Spot
     Spot Image
EBU (European Broadcasting Union), 42
Ecole Polytechnique, 62
ECS (European Communications Satellite), 43–5, 54, 55, 61, 65, 102, 111
Edwards Air Force Base, 81, 85, 133
East Germany,
  see Germany, East
Eiffel Tower, 198–9
ELA-1/2, 57–8
  see Kourou

ELDO (European Launcher Development Organisation), 9, 10–12, 15–22, 35, 39, 42
ELDO-B rocket, 12
Emeraude, 13–14
Energia, 64
Engel, Rolf, 6
Engstrom, Fredrik, 174
Eosat (Earth Observing Satellite Co), 91, 93
EPrime Aerospace Corp, 199
ERA, 137–8
Eridan, 124
ERNO, 43
    see MBB
ERS (Earth Resources Satellite), 61, 101, 142-4
ESA (European Space Agency), 21–2, 25–32, 39–41, 43, 44, 46, 47, 51, 52, 54, 61, 65, 66, 74–8, 81–4, 94–6, 98, 101, 106, 123, 142, 144–8, 153–4, 157, 160–76, 181–2, 185, 191
ESDAC (European satellite tracking and data processing centre), 10
Esnault-Pelterie, Robert, 5–6
ESOC (European Space Operations Centre), 41, 47, 95–6
Espionage, 62–4
ESRIN (European Space Research Institute), 10
ESRO (European Space Research Organisation), 9–10, 18, 21, 22, 33–41, 42, 70
ESRO satellites, 33, 35–9
ESTEC (European Space Technology Centre), 10, 33, 35, 36, 40
ETCA (Etudes Techniques et Constructions Aérospatiales), 23, 40, 112, 148, 158
Eumetsat, 48
Eureca (European Retrievable Carrier), 172
Europa rocket, 9, 10–13, 15– 22, 26, 42
European Community, 90–1, 109, 122
European Space Conference, 15, 21, 42, 70
European Space Programmes
    see Ariane
        Columbus

COSARI
COS-B
ECS
ELDO-B
ERS
ESRO satellites
Europa
Giotto
GEOS
HEOS
Hermes
Hipparcos
International Halley Watch
International Space Station
L-3S
LAS
MARECS
Meteosat
OTS
Spacelab
TD-1/2
European Television Forum, 121
Eurosatellite, 112
Eutelsat (European Telecommunications Satellite Organisation), 45, 47, 61, 65, 102–3, 107, 111, 116–18, 146, 187
EVA (Extravehicular Activity)
    see Space Walk

Fabius, Laurent, 114
Falklands campaign, 105, 149, 152
FAO (United Nations Food and Agricultural Organisation), 143
Ferranti, 40
Feustel-Buechl, Jorg, 66
Fiat
    see Snia BPD
Filmnet, 111
Financial Times, 111, 166
Finland, 98, 121
    see Tele-X
Firewheel, 51
Fletcher, James, 61, 141
Fleury, Michel, 62
Fleya, 195
Fluid Gravity, 148
Fokker, 40, 95, 158
Food,
    see Space food
Ford Aerospace, 107, 131

France, 2, 5–7, 9, 11–21, 23, 25, 27,
    30, 40, 42–6, 49–50, 54–7,
    62–5, 70, 87–95, 97–101, 103–5,
    111–17, 122, 123– 40, 144–6,
    149, 151–2, 154–8, 161, 163,
    167, 169–70, 174, 176, 179–80,
    182, 191, 197
  see Astérix
    Auréol
    CNES
    Coralie
    Diamant
    Dragon
    Emeraude
    Eridan
    Phobos
    Saphir
    Signe
    Spot
    SRET
    Symphonie
    TDF
    Télécom
    Topaze
    Véronique
France Télécom, 66
Freedom,
  see International Space Station
Frog, 98
Fullerton, Gordon, 82
Furrer, Reinhard, 83–4, 174

Gagarin, Yuri, 10
Gamma-ray astronomy, 34, 40, 126
Garriot, Owen, 76–8
GE Astro, 113, 120
Gemini, 13, 175
General Dynamics, 68, 188
  see Atlas Centaur
General Electric, 106
GEOS, 26, 40–1
Gerasimov, Gennady, 62
Germany, East, 97, 127
Germany, West, 1–2, 4–5, 9, 11–12,
    16–18, 20–1, 23, 25, 40, 42–3,
    45–6, 51, 54, 70, 83–5, 95,
    100–1, 107, 111–17, 122, 146,
    149–50, 153–4, 158, 161, 163,
    167, 169, 174, 179–80, 183, 191
  see Astris
    D1 Spacelab mission
    DFS Kopernikus
    DFVLR

    Firewheel
    Saenger
    Symphonie
    TVSAT
Giacobini-Zinner, 95
Gibson, Roy, 41, 161, 164, 176, 177
Giotto, 55, 94–7
Giscard d'Estaing, Valéry, 29, 121
Glavkosmos, 68
Glenn, John, 10
GMS, 194
GOES satellites, 46, 60, 92
Goldsmith, P., 47, 101
Gorbachev, Mikhail, 136, 139, 192
Goumy, Claude, 104
GPS (Global Positioning System)
    satellites, 61
Granada Television, 119
Greenpeace,
  see Rainbow Warrior
Grigg-Skjellerup, 97
GRU (Soviet military intelligence
    service), 63
GSOC (German Space Operations
    Centre), 83
  see Oberpfaffenhofen
GSTAR, 55, 58
GTE Spacenet, 58, 66, 105

H-I/II rocket, 68, 194–5
H8 engine, 28, 53
H20 engine, 18
Hague, The, ministerial meeting,
    143, 147, 162, 165–73, 176, 183
Halpern, Richard, 162
Hammaguir, 6, 14
Halley, Edmund, 94
Halley's comet, 55, 94–7
  see Giotto
Harrington, Lou, 178
Hartsfield, Hank, 84
Haury, Jean-Michel, 62
Hauzeur, M.-A., 28
Hawker-Siddeley Dynamics, 43
HDRR (High-Data-Rate
    Recorder), 79
Helios, 87, 91, 151–2
HEOS satellites, 33, 35–9
Hermes, 154–9, 163–77, 179, 181–2,
    184, 194
Heyss Island, 123
Hipparcos, 67, 145
Hitler, Adolf, 1

HM-7 engine, 24–5, 27, 53, 60
Hocker, Alexander, 39
Honeywell, 40
Horus, 179–81, 185
Hotol, 179, 181–4, 185
House of Commons, British, 164–5, 170
House of Lords, British, 197–8
Houston Chronicle, 99
Hughes Aircraft Co, 131
Hungary, 97, 98, 127
Huygens, 148
Hypersonic flight, 177–85

IAF (International Astronautical Federation), 165
ICE (International Cometry Explorer), 94–5
IGN (French National Geographical Institute), 89
IL-76, 127
Independent, 119, 165
India, 26, 52, 100, 196
  see Apple
    Insat
    ISRO
Inmarsat (International Maritime Satellite Organisation), 52, 61, 107–9
Insat, 61
INSEE (French national statistics institute), 62
Insurance, 56–7, 185–8
Intelsat (International Telecommunications Satellite Organisation), 27, 28, 42, 59, 61, 66, 107–8, 187
Intelsat satellites, 54, 59, 66, 67, 85, 107–8
Inter-Agency Consultative Group, 95
Intercosmos, 127
International Halley Watch, 94–7, 147
  see Giotto
International Herald Tribune, 108
International Magnetospheric Study, 41
International Space Station, 86, 101, 153–4, 159–75, 185, 193
  see Columbus
    Hermes
    Polar platform
Ionosphere, 38, 124

Ireland, 98, 170
Iris satellite
  see ESRO-II
ISAS (Japanese Institute of Space and Astronautical Science), 94
ISEE-3,
  see ICE
ISF (Industrial Space Facility), 172–3
ISO (Infrared Space Observatory), 145–6
Israel, 196
  see Ofek-1
    Shavit
ISRO (Indian Space Research Organisation), 52
Istres, 156
Italsat, 107
Italy, 9, 11, 16, 40, 43, 45–6, 48, 53, 70, 95, 100–1, 107, 145, 154, 158, 163, 169, 189
  see Italsat
    Scout
    Sirio
Ivanchenkov, Alexander Sergeyevich, 123, 129–30

JAL (Japan Air Lines), 108
Japan, 46, 61, 68, 86, 92, 94– 5, 100, 120, 140, 142, 153–4, 160–2, 167, 183–5, 193–5
  see H-1/2
    International Space Station
    ISAS
    MOS
    N-I/II
    NASDA
    Sakigake
    Suisei
Johns Hopkins ultraviolet telescope, 86
Johnson Space Center, 85
Jupiter control centre,
  see Kourou
Jupiter, 147
Juppé, Alain, 115

Kaminski, Heinz, 20
Kennedy, John F., 12, 153
Kerguelen, 123, 125
KH-11, 60
Kiruna, 88, 124
  see Spot Image

Kizim, Leonid, 129
Kohl, Helmut, 80, 149, 167
Komsomolskaya Pravda, 139
Konorev, Valery, 63
Kourou, 19, 26–32, 49–59, 65–6, 87, 95, 154, 156
Krikalyov, Sergei, 137
Kvant, 135

L-3S, 21–2
Laben, 40, 95
La Cinq, 104
L'Air Liquide, 23
Lambton, Michael, 77
Land mobile communications, 109
Landsat, 87–9, 91–4, 194
La Repubblica, 99
LAS (Large Astronomical Satellite), 33
Laser, 124, 131–2
L'Astronautique, 6
Lavany, Michel, 61, 68
Lawson, Mark, 119
Lebeau, A., 46
Lebedev, Valentin Vitalyevich, 128
Le Courrier de Mantes, 63
Le Figaro, 62, 65, 115
Le Monde, 63, 152
Leonov, Alexei, 13
Le Point, 114
Les Echos, 158
Leskov, Sergei, 139
Les Mureaux,
  see Aerospatiale
Libération, 65
Lichtenberg, Byron, 76
Life sciences, 78, 80, 82–5, 125, 129–30, 172
Lifestyle, 111
LittLEO, 189, 199
Lockheed, 82
Long March, 68, 188, 195–6
Longuet, Gerard, 114
LTV, 189
Lucid, Shannon, 132
Luest, Reimar, 162, 165, 169, 182
Lunik, 7, 124
Lunokhod, 124
Luxembourg, 113
  see Astra
Luyken, Georg-Michael, 121

Lyakhov, Vladimir
  Afanasevich, 125, 139
Lyons, Jacques-Louis, 171

M6, 104
Macey, Air Vice-Marshal E. H., 106
Madelin, Alain, 170
Magnetic fields,
  Interplanetary, 33
Magnetosphere, 36, 37, 38, 41, 124
Maillard, Philippe, 62
Maine, Steve, 118
Mallet, E. S., 44
Manarov, Musa, 135, 137, 139
Manned space flight, 1, 10, 17, 19, 70, 76–9, 81–5, 123–39, 147, 153–88
  see Apollo
    Buran
    Columbus
    Gemini
    Hermes
    International Space Station
    Kvant
    Mir
    Salyut
    Skylab
    Soyuz
    Spacelab
    Space shuttle
    Voskhod
    Vostok
Manole, Antonetta, 62
Marconi, 40, 105–6, 148
MARECS, 44, 52, 53, 55, 108
Mariner, 97
Marisat, 108
Mars, 97–100, 124, 191–3, 198
  see Phobos
Martinand, Claude, 89
Martin, Marietta, 60, 68, 107, 188
  see Titan
Mason, Brian, 101
Matagorda Island, 199
Matra, 23, 43, 70, 87, 101, 103, 145, 158, 196
Max Planck Institute, 40
MBB (Messerschmitt-Boelkow-Blohm), 18, 25, 40, 67, 70–1, 74, 101, 107, 112, 146, 158, 172, 179
McCaskill, Allan, 187

McDonnell Douglas, 60, 68, 178, 188, 193
  see Delta
Medical experiments, 83–5, 129–33
Merbold, Ulf, 1, 76–7, 83, 123, 174
Mercury, 175
MESH consortium, 43
Messerschmid, Ernst, 83–4, 155, 174
Meteorological Office, British, 100
Meteorological research, 26, 45–8, 52, 66, 100–1, 123–4, 194
  see ERS
    GMS
    Meteorological Office, British
    Meteosat
    Metsat
Meteosat, 26, 45–8, 52, 66, 100–1, 143–4
Metsat, 92
Mexico, 131
Meyrin conference, 8
Microgravity, 80, 82, 85
Milky Way, 133
Ministry of Defence, British, 106, 181, 184
Mir, 64, 78, 134–5, 137–8
Mission to Planet Earth, 141–2
MIT (Massachusetts Institute of Technology), 178
  see Technology Review
Mitsubishi, 184
Mitterrand, François, 55, 129, 135–6, 180
Mohmand, Abdul Ahad, 139
MOM (US Mars Orbiter Mission), 100
Mongolia, 127
Moon, 7–8, 17, 76, 124, 159, 191
MOP, 100–1
Morelos, 131
MOS, 194
MR-12 rockets, 123
MTFF (Man-Tended Free Flyer), 172–3
Mulley, Fred, 15
Murdoch, Rupert, 111, 118–20
Mururoa,
  see Rainbow Warrior

N-I/II, 193–4
Nagel, Steven, 84
NASA (National Aeronautics and Space Administration), 10, 12, 17, 18–19, 20, 33, 36, 38, 40, 41, 47, 60, 70, 74–7, 79–83, 85–6, 130, 132, 141, 144, 146–8, 153, 159– 62, 173–4, 177–8, 185–6, 191–2, 199
NASDA (Japanese national space agency), 142
Natal tracking station, 31–2, 53, 193
NASP (US National Aerospace Plane), 177, 185
NATO (North Atlantic Treaty Organisation), 44, 105–6, 149
NATO satellite, 44
NBC, 119
Neeman, Yuval, 197
Netherlands, The, 9, 16–17, 40, 71, 95, 111, 158, 170, 174
New York Times, 64
Nicolaev, Anatoliy, 92
Nicollier, Claude, 174, 176
Nimbus, 144
NOAA (US National Oceanic and Atmospheric Administration), 91–2
Noordwijk,
  see ESTEC
Nordhausen, 5
Nordiska Satellitaktiebolaget, 121
Northern lights,
  see Aurora Borealis
Norway, 121, 170, 189, 199
  see Tele-X
Novikov, Yulian, 93

Oberpfaffenhofen, 83, 85
Oberth, Hermann, 4
Ockels, Wubbo, 77, 83–5, 174–5
Ofek, 196
Olympus, 146
OMB (US Office of Management and Budget), 159
OMEGA, 123
ONERA (Office National d'Etudes et de Recherches Aeronautiques), 6
Orye, R., 27
Oscar, 51
OTS (Orbital Test Satellite), 26, 43–5, 46, 106, 118, 146
Overmyer, Robert, 82
Page, D. E., 36
PAL, 117, 119
Pan Am Pacific, 195

Pan Am Sat, 66
Pardoe, Geoffrey, 8
Parker, Robert, 76, 78–9
Pattie, Sir Geoffrey, 165, 182, 183
Pearson, 119
Peenemuende, 4–5, 6
Pentagon, 150, 159–60, 162, 173, 178
Perseus, 133
Phobos, 97–100
Pils, Wolfgang, 6
Poland, 98, 127
Polar platform, 101, 143, 168, 172–3
Pollution, 80, 87–8, 83, 141–2
Polyakov, Valery, 135
Pompidou, Georges, 20
Post Office, British, 45, 112
Probes, 7, 55, 94–100, 124–5, 147–8
  see Cassini
    Giotto
    Huygens
    ICE
    Lunik
    Mariner
    Mars
    Phobos
    Sakigake
    SOHO-CLUSTER
    Suisei
    Vega
    Venera
    Viking
Proton, 68–9, 188
Pyke, Thomas, 92

QE2, 108
Quayle, Dan, 192
Quistgaard, Erik, 77, 79, 81

RAI, 111
Rainbow Warrior, 55
Rasse, P., 24
RCA, 113
Reagan, Ronald, 68, 80, 130–1,
    149–50, 153–4, 159–60, 173,
    178, 195, 198
Remek, Valdimir, 127
Remote sensing
  see Earth observation
Ride, Sally, 141
  see Ride report
Ride report, 141–2, 191–3
Riesenhuber, Heinz, 158–9, 167,
    170, 173

Ripoll, Andres, 174
Rippon of Hexham, Lord, 7, 16
RITA, 103
Rocard, Michel, 117
Rockets,
  see A1-4 series
    Ariane
    Atlas Centaur
    Blue Streak
    Coralie
    Delta
    Diamant
    Dragon
    ELDO- B
    Emeraude
    Europa
    H-1/2
    Long March
    Lunik
    MR- 12
    Proton
    Saphir
    Saturn
    Scout
    Skylark
    Thor-Delta
    Titan
    Topaze
    V1-2 series
    Véronique
Rockwell International, 178
Rogers, William P.,
  see Rogers Commission
Rogers Commission, 192
Rolls-Royce, 11, 182–4
Romanenko, Yuri, 64
Romania, 127
Rome, ministerial meeting, 153, 157
Roulet, Marcel, 104
Royal Aircraft Establishment, 7, 184
Ryumin, Valeri Viktorovich, 125

Saab, 43
Saab-Scania, 43
Saclay nuclear study centre, 40
Saenger, 179–81, 184–5
  see Cargus
    Horus
Saenger, Eugen, 180
Safety, 155– 6
Sakigake, 94–5
Salyut, 72–3, 77, 78, 125, 127–30
San Marco, 189

Saphir, 13–14
SAR (Synthetic Aperture Radar), 142
Sarkssian, Irina, 62
Sat 1, 111
Satellites,
  meteorological, 45–8, 100–1
  military, 44, 60–1, 87, 151–2
  remote sensing 87–94, 101, 141–4, 151–2
  scientific, 33– 41
  telecommunications, 42–5, 102–9, 146–7
  television, 110–22
Saturn, 147–8
Saturn rocket, 17, 73
Savinykh, Viktor, 133
Scout, 33, 36, 189
Screensport, 111
SDI (Strategic Defense Initiative), 86, 130–1, 149–51, 160
SDIO (Strategic Defense Initiative Organization), 131
Seasat, 144
Selenia, 40, 43, 101, 107
SENER Ingenieria y Sistemas, 148, 158
SEP (Société Européenne de Propulsion), 18, 23, 25, 27–8, 60, 62–3, 95, 148
SEREB (Société pour l'Etude et la Réalisation d'Engins Balistiques), 13
SES (Société Européenne des Satellites), 113, 117–18, 120
  see Astra
Shackleton, Lord, 198
Shavit, 196
Shaw, Brewster, 76, 78
Shevardnadze, Eduard, 136
Shultz, George, P., 173
Shuttle,
  see Buran
    Hermes
    Space Shuttle, US
Signe, 125
Simonds-Gooding, Anthony, 120
Sirio, 48, 53
Skylab, 73–4, 76, 148, 175
Skylark, 7
Skynet, 44, 61, 105–7
Sky Television, 111, 118–20, 122
Slayton, Deke, 199

Smith, W. H., 111, 119
SMS-1 satellite, 46
Snia BPD, 67, 189
Société Française de Physique, 5
SOHO, 148
Solar observation, 82, 86
Solar radiation, 37, 124
Solar wind, 33, 148
Solovyov, Vladimir, 129
Soviet Union, 1, 7, 10, 12–13, 17, 46, 62–4, 68–9, 72–3, 78, 91–100, 123–30, 132–40, 151. 188, 191–3, 196–7
  see Auréol
    Buran
    Cosmos
    Energia
    International Halley Watch
    Lunik
    Lunokhod
    Mars
    Mir
    Phobos
    Salyut
    Soyuz
    Soyuzkarta
    Sputnik
    Vega
    Venera
    Voskhod
    Vostok
Soyuz, 1, 73, 77, 123, 125, 128–9, 137, 139
Soyuzkarta, 92–3
Space debris, 61–2
Space food, 138–9
Space Industries Inc, 172
Spacelab, 1, 19, 20, 21, 22, 70–86, 123, 155, 172, 174
Spacenet, 54, 55, 58, 66, 105
  see GTE Spacenet
Space Services Inc, 199
Space shuttle, US, 1, 18–19, 22, 68, 71–2, 74–86, 128, 130–3, 155–6, 159, 172–3, 175–6, 187–9, 192–5
  see Atlantis
    Challenger
    Columbia
    Discovery
Space station
  see International Space Station
Space tug, 18–19
Space walk, 13, 137–8

Spain, 9, 40, 71, 158, 170
Spartan, 133
Spot, 57, 61, 87–94, 142, 151
  see Helios
    Landsat
    Spot Image
Spot Image, 88–9, 91
Sputnik, 6–7
Spycatcher, 7
SRET, 124
Star City, 127–8, 134–5
Star Wars
  see SDI
State Department, US, 159, 162
State of the Union address, 159
Stoltenberg, Gerhard, 158
Strub, Herman, 113
Sugar, Alan, 118, 120
Suisei, 94–5
Superchannel, 111
Sweden, 9, 43–4, 46, 87, 98, 121,
    161, 170, 195
  see Fleya
    Tele-X
Switzerland, 9, 45–6, 71, 98,
    161, 170, 174
Symphonie, 42–3
Syracuse, 103

Tanegashima, 193
Tass, 7
TD-1/2 satellites, 33, 37–9
TDF (Télédiffusion de France), 112,
    114–15
TDF satellites, 59, 61, 112–17,
    121–2
  see TVSAT
TDRS (Tracking and Data Relay
    Satellite), 75, 79, 85
Technology Review, 178
Télécom satellites, 44, 55, 66, 103–4,
    116
Television, 102, 104, 110–22, 135–6,
    140
  see CNN
    D2-MAC Packet
    Disney
    Filmnet
    La Cinq
    Lifestyle
    M6
    PAL
    RAI

Sat 1
Satellite television
Screensport
Sky
Superchannel
TV 5
Tele-X, 121
Telstar, 42, 131
Terma Electronik-Centralen, 40
Terra-Mar Resource
    Information Services, 93
Thames Television, 113
Thatcher, Margaret, 149, 158, 164,
    182, 197
Thomson-CSF, 95, 103
Thor-Delta, 37, 40, 41, 193
Thornton, Bill, 81
Three Mile Island, 141
Time magazine, 192
Times, The, 19, 165–6
Titan, 147–8
Titan rocket, 60, 68, 107, 187–8, 196
Titov, Vladimir, 135, 137, 139
Tognini, Michel, 135
Topaze, 13–14
Topex-Poseidon, 144, 151
Toulouse, 87–8, 156
  see Hermes
    Spot Image
Treasury, British, 164
TSW, 113
TV 5, 111
TVSAT, 59, 66, 112–17, 121–2
  see TDF
Tyuratam, 88

Ulster Television, 113
Ultra-violet astronomy, 34, 38, 86
United States, 1, 10, 12–13, 17–19,
    22, 26–7, 40, 46, 55, 58, 60–2,
    66–9, 70–86, 87, 91–4, 100, 108,
    123, 127–8, 130–4, 141–2, 144,
    146–9, 151, 153–4, 159–63,
    166–8, 170–3, 177–9, 183–6,
    188–9, 191–3, 197–8
  see Apollo
    Atlas Centaur
    Big Bird
    Gemini
    GOES
    GPS
    GSTAR
    ICE

International Halley Watch
International Space Station
KH- 11
Landsat
Mariner
Marisat
Metsat
MOM
NASA
Saturn
Skylab
SMS-1
Spacenet
Space Shuttle
Thor-Delta
Titan
Viking

V1-2 rockets, 5, 180
Van Allen belts, 35
Van de Hulst, Professor Henk, 37
Vandenberg Air Force Base, 60, 189
Varygina, Lyudmila, 62
Vega, 94–7
Venera, 125
Venus, 125
Verdier, Pierre, 62–3
Verein fuer Raumschiffahrt, 4
Vernon, 5, 6, 25, 27–8, 60, 62–3
Véronique, 6, 49
Vietnam, 127
Vignelles, Roger, 24, 66
Viking engine, 17, 24–5, 27, 30, 31,
    51
Viking probes, 97
Virgin, 111

Volkov, Alexander, 137–8
Von Braun, Werner, 1, 4–5, 180
Voskhod, 13
Vostok, 10

Walklet, Don, 93
Wallops Island, 189
Walter, Paul, 62
Waylan, Gerry, 105
Weinberger, Caspar, 149, 195
Western Test Range, 33, 36
Western Union Corp, 195
West Germany,
    see Germany, West
Westinghouse Electric Co, 172
White, Ed, 13
White Sands, 85
Whitten, Patrick, 112
Whittle, Frank, 184
Williams, Charles, 93
WMO (World Meteorological
    Organisation), 100
Woomera, 7, 9, 12, 18
World Weather Watch, 100
Wright, Pearce, 19
Wright, Peter, 7

X-15/30, 178
Xichang, 195
X-ray astronomy, 34, 35, 82, 124,
    126, 130

Young, John, 76, 78, 81
Young, Lord, 197

Zircon, 151